Springer Theses

Recognizing Outstanding Ph.D. Research

For further volumes:
http://www.springer.com/series/8790

Aims and Scope

The series "Springer Theses" brings together a selection of the very best Ph.D. theses from around the world and across the physical sciences. Nominated and endorsed by two recognized specialists, each published volume has been selected for its scientific excellence and the high impact of its contents for the pertinent field of research. For greater accessibility to non-specialists, the published versions include an extended introduction, as well as a foreword by the student's supervisor explaining the special relevance of the work for the field. As a whole, the series will provide a valuable resource both for newcomers to the research fields described, and for other scientists seeking detailed background information on special questions. Finally, it provides an accredited documentation of the valuable contributions made by today's younger generation of scientists.

Theses are accepted into the series by invited nomination only and must fulfill all of the following criteria

- They must be written in good English.
- The topic should fall within the confines of Chemistry, Physics, Earth Sciences, Engineering and related interdisciplinary fields such as Materials, Nanoscience, Chemical Engineering, Complex Systems and Biophysics.
- The work reported in the thesis must represent a significant scientific advance.
- If the thesis includes previously published material, permission to reproduce this must be gained from the respective copyright holder.
- They must have been examined and passed during the 12 months prior to nomination.
- Each thesis should include a foreword by the supervisor outlining the significance of its content.
- The theses should have a clearly defined structure including an introduction accessible to scientists not expert in that particular field.

Pei Zhang

Beam Diagnostics in Superconducting Accelerating Cavities

The Extraction of Transverse Beam Position from Beam-Excited Higher Order Modes

Doctoral Thesis accepted by
the University of Manchester, UK

Author
Dr. Pei Zhang
BE-RF
CERN
Geneva
Switzerland

Supervisors
Prof. Dr. Roger M. Jones
The Cockcroft Institute
School of Physics and Astronomy
The University of Manchester
Manchester
UK

Dr. Nicoleta Baboi
Machine Diagnostics and
Instrumentation Group
Deutsches Elektronen-Synchrotron
Hamburg
Germany

ISSN 2190-5053 ISSN 2190-5061 (electronic)
ISBN 978-3-319-34616-8 ISBN 978-3-319-00759-5 (eBook)
DOI 10.1007/978-3-319-00759-5
Springer Cham Heidelberg New York Dordrecht London

Softcover reprint of the hardcover 1st edition 2013

Printed on acid-free paper

Springer is part of Springer Science+Business Media (www.springer.com)

Supervisors' Foreword

The introduction of a charged particle beam into an accelerating cavity self-excites a wakefield which acts back on the bunch itself (the short-range portion) and on succeeding bunches (the long-range portion) in a train of accelerated particles. This wakefield has both transverse and longitudinal components. This work is focused on the transverse component in the main and on its implications on the long-range component in particular. This transverse wakefield can be decomposed into a series of multipoles and, providing the beam is guided close to the electrical centre of the cavity, the dipole mode is the dominant one. If the progress of this wakefield is left unchecked, it can appreciably dilute the beam's quality. Higher order mode (HOM) couplers are incorporated into the cavity to suppress the wakefields. In this work, the wakefield is also sampled in order to provide a remote diagnostic as to the beam position within the cavity—the amplitude of the dipole modes excited being directly proportional to the offset of the beam.

This technique has previously been exploited for superconducting accelerating cavities operating at 1.3 GHz by colleagues from CEA, DESY, SLAC, and FNAL. However, this work goes further in addressing the application of the method to third harmonic cavities for bunch shaping light source applications. Here, there are four superconducting cavities within a module—and almost all modes are coupled throughout the module. This is a challenging problem to address both computationally, and to practically implement as a diagnostic. Nonetheless, substantial progress is indicated in this text to address both issues. Bands of modes rather than individual modes are found to be particularly effective. Experimental results indicate a position resolution, based on this technique, of 20 microns over a complete module of four cavities. Means are also indicated to improve the resolution and to apply this method to other cavity configurations.

This research is distinguished by its clarity and potential for application to several other international facilities, using superconducting accelerating cavities. The material is pedagogically appropriate, and indeed highly recommended, for both students new to the field, and to scientists familiar with the fields of accelerating cavities and beam diagnostics.

Manchester, April 2013 Prof. Dr. Roger M. Jones
Hamburg Dr. Nicoleta Baboi

Acknowledgments

First of all, I would like to thank my supervisors, Prof. Roger M. Jones and Dr. Nicoleta Baboi, for their encouragement and constant support over the last 3 years. They introduced me to the field of accelerator physics and guided me throughout my Ph.D. I am particularly grateful to Nicoleta for her patience and immeasurable support from all aspects of my life at DESY. The HOMBPM project could never stand in the current strong position without her passion, experience, coordination, and strong involvement. Coming from a high energy physics background, my achievements in accelerator physics today would certainly not be possible without Roger's excellent tutorship.

Based at DESY for the last 3 years, I thank all my colleagues in Hamburg for their support and guidance, especially Dirk Lipka, Bastian Lorbeer, and Thomas Wamsat. I also thank Kay Wittenburg and Dirk Nölle for supporting my research at DESY from many aspects. I extend my gratitude to colleagues in MDI group for helping me with HOM measurements at FLASH, namely Klaus Knaack, Reinhard Neumann and Norbert Wentowski. Inspiring remarks from colleagues in MPY group deserve a special mention, namely Martin Dohlus and Rainer Wanzenberg. I thank Alexandr Ignatenko for his company when we shared an office together in Building 30 at DESY. I thank all members of Manchester Electrodynamics and Wakefields (MEW) group for discussions on the weekly meetings. I particularly thank Ian Shinton for his numerous help and wise advice. I extend my thanks to collaborators in Rostock University and FNAL, namely Nathan Eddy, Thomas Flisgen, Hans-Walter Glock, and Manfred Wendt. Our discussions were always constructive and our collaborations were both successful and pleasant.

I acknowledge the funding support of EuCARD, DESY, The University of Manchester and Cockcroft Institute. During my Ph.D., I received bursaries to attend three international conferences: LINAC2010, IPAC2011 and DITANET conference on Beam Diagnostics, and financial support for two international workshops: DIPAC2011 and HOMSC12, I would like to thank the respective organization committees and institutes. I also appreciate the opportunities to attend two CAS, one JUAS, one LC, and one DITANET school. My research was benefited a lot from these conferences and schools.

Finally, I would like to express my thanks to my parents for their love and support. My Mum transmitted me the strength to pursue my dreams, while my Dad instilled in me the curiosity for all things in nature. Last but not least, I thank my wife, Xi Wang, for her love and company.

Contents

1 Introduction 1
- 1.1 Free-Electron Laser in Hamburg 2
- 1.2 Third Harmonic Cavities 3
- 1.3 Wakefields 8
  - 1.3.1 Electromagnetic Field in a Cavity 9
  - 1.3.2 Definitions of RF Parameters 10
  - 1.3.3 Definition of Wakefields 11
- References 14

2 Electromagnetic Eigenmode Simulations of the Third Harmonic Cavity 17
- 2.1 The Third Harmonic Cavity as a Periodic Structure 17
- 2.2 The Beam Pipe as a Circular Waveguide 22
- 2.3 Eigenmodes of an Ideal Third Harmonic Cavity 25
- References 30

3 Measurements of HOM Spectra 31
- 3.1 Transmission Spectra of an Isolated Single Cavity 31
- 3.2 Module-Based Transmission Spectra 35
- 3.3 Beam-Excited HOM Spectra 37
- References 42

4 Analysis Methods for Beam Position Extraction from HOM 43
- 4.1 Data Preparation 43
- 4.2 Direct Linear Regression 45
- 4.3 Singular Value Decomposition 48
- 4.4 *k*-means Clustering 54
- 4.5 Comparison of DLR, SVD and *k*-means Clustering 54
  - 4.5.1 Fixed Sample Split 54
  - 4.5.2 Cross-Validation 56
- References 60

5 Dependencies of HOM on Transverse Beam Offsets 61
5.1 Measurement Scheme 61
5.2 The Localized Dipole Beam-Pipe Modes 62
5.3 Trapped Cavity Modes in the Fifth Dipole Band. 63
5.4 Coupled Cavity Modes in the First and the Second Dipole Band 67
References 70

6 HOM-Based Beam Position Diagnostics 71
6.1 The Principle of Custom-Built Test Electronics. 71
6.2 Position Diagnostics with Trapped Cavity Modes 73
6.2.1 Data Preparation 74
6.2.2 Extraction of Beam Position from HOM Waveforms. . . . 76
6.2.3 Cross-Validation 76
6.2.4 The Search for Suitable Trapped Dipole Modes 79
6.2.5 The Search for a Suitable Time Window 81
6.2.6 Results for all HOM Couplers 84
6.3 Position Diagnostics Based on a Band of Coupled Modes 84
6.4 Fundamental Limitations to Position Resolution 88
References 90

7 Conclusions 91
7.1 Summary 91
7.2 Future Studies on HOMBPM 93
References 93

Appendix A: Mathematics 95

Appendix B: Eigenmodes of an Ideal Third Harmonic Cavity 101

Appendix C: Technical Details of the HOM Measurements. 115

About the Author 117

Abstract

Higher order modes (HOM) are electromagnetic resonant fields. They can be excited by an electron beam entering an accelerating cavity, and constitute a component of the wakefield. This wakefield has the potential to dilute the beam quality and, in the worst case, result in a beam-break-up instability. It is therefore important to ensure that these fields are well suppressed by extracting energy through special couplers. In addition, the effect of the transverse wakefield can be reduced by aligning the beam on the cavity axis. This is due to their strength depending on the transverse offset of the excitation beam. For suitably small offsets the dominant components of the transverse wakefield are dipole modes, with a linear dependence on the transverse offset of the excitation bunch. This fact enables the transverse beam position inside the cavity to be determined by measuring the dipole modes extracted from the couplers, similar to a cavity beam position monitor (BPM), but requires no additional vacuum instrumentation.

At the FLASH facility in DESY, 1.3 GHz (known as TESLA) and 3.9 GHz (third harmonic) cavities are installed. Wakefields in 3.9 GHz cavities are significantly larger than in the 1.3 GHz cavities. It is therefore important to mitigate the adverse effects of HOMs to the beam by aligning the beam on the electric axis of the cavities. This alignment requires an accurate beam position diagnostics inside the 3.9 GHz cavities. It is this aspect that is focused on in this thesis. Although the principle of beam diagnostics with HOM has been demonstrated on 1.3 GHz cavities, the realization in 3.9 GHz cavities is considerably more challenging. This is due to the dense HOM spectrum and the relatively strong coupling of most HOMs amongst the four cavities in the third harmonic cryo-module.

A comprehensive series of simulations and HOM spectra measurements have been performed in order to study the modal band structure of the 3.9 GHz cavities. The dependencies of various dipole modes on the offset of the excitation beam were subsequently studied using a spectrum analyzer. Various data analysis methods were used: modal identification, direct linear regression, singular value decomposition, and k-means clustering. These studies lead to three modal options promising for beam position diagnostics, upon which a set of test electronics has been built. The experiments with these electronics suggest a resolution of 50 micron accuracy in predicting local beam position in the cavity and a global resolution of 20 micron over the complete module.

This constitutes the first demonstration of HOM-based beam diagnostics in a third harmonic 3.9 GHz superconducting cavity module. These studies have finalized the design of the online HOM-BPM for 3.9 GHz cavities at FLASH.

Keywords Higher order mode · Wakefield · Beam position diagnostics · Superconducting accelerating cavity · Linear regression · Singular value decomposition · *k*-Means clustering · Cross-validation

Chapter 1
Introduction

A charged particle bunch interacts electromagnetically with the accelerating cavities in an accelerator and subsequently generates an electromagnetic field, namely the wakefield [1]. There are short range wakefields with effects within a bunch, and long range wakefields lingering after the bunch traverses the cavity, and acts back on the following bunches entering the cavity and perturbs their motion. The wakefield has both longitudinal and transverse components. The former performs a longitudinal perturbation and may consequently change the energy spread of the following bunch. The latter kicks the following bunch transversely and therefore may cause a growth of the multi-bunch transverse emittance. Both effects deteriorate the beam quality and in the worst scenario result in a beam-break-up instability [2]. It is therefore important to ensure that these fields are well suppressed. Indeed, for the acceleration of high intensity particle beams, suppressing wakefields in both superconducting [3] and normal conducting [4] cavities is a common requirement.

The wakefield can be expanded as a multipole series and each term is a so-called mode. In the cavity with a cylindrical symmetry, one can further categorize the modes into monopole, dipole, quadrupole, etc. modes according to their azimuthal field distributions [5] and these modes are grouped in bands. Modes have higher frequencies than the accelerating mode are often called higher order modes (HOM).

Wakefields can be suppressed by either aligning the beam carefully to mitigate the excitation of certain HOMs, or by extracting HOM energies out of the cavities, which is often realized with specially designed couplers. In many practical cases, the longitudinal wakefield is dominated by monopole modes, while the transverse one by dipole modes. The dipole transverse wakefield is of particular interest, since it has a linear dependence on the transverse offset of the excitation beam, regardless of the position where one examines it. In other words, beam-excited dipole modes can be used to provide direct information about the transverse beam position inside the cavity, therefore a HOM beam position monitor (HOMBPM).

FLASH [6] is a free-electron laser user facility driven by a superconducting linear accelerator (linac). There are two types of accelerating cavities in FLASH linac: TESLA 1.3 GHz cavities [7, 8] and 3.9 GHz cavities [9]. HOMBPMs have been

P. Zhang, *Beam Diagnostics in Superconducting Accelerating Cavities*,
Springer Theses, DOI: 10.1007/978-3-319-00759-5_1,

previously built for 1.3 GHz cavities and it is planned to build similar system for 3.9 GHz ones. However, significant challenges arise due to the prominent features of the 3.9 GHz cavities. This thesis focuses on the feasibility study of beam position diagnostics with HOMs in 3.9 GHz cavities and consequently defined the specifications of the corresponding HOMBPM system.

In this chapter, the FLASH facility and the third harmonic cavities are first described followed by a review of the basic concepts related to wakefields and the principle of diagnostics using dipole modes. RF parameters used in this thesis are also defined. Chapter 2 proceeds with eigenmode simulations of an ideal third harmonic cavity without couplers. The modal structure of the cavity is characterized and field distributions of various HOMs are presented. Experimental investigations of HOMs in third harmonic cavities are summarized in Chap. 3; this brings together the RF and beam-based-HOM measurements with the results of eigenmode simulations described in Chap. 2, depicting a complete picture of HOMs in these cavities and pointing out the options for diagnostics: localized and multi-cavity modes. Prior to describing the dependencies of dipole modes on transverse beam offsets, statistical methods used to extract beam position information from HOM signals are presented in Chap. 4. Three different methods are elucidated in depth: direct linear regression, singular value decomposition and k-means clustering. These are compared in terms of performance and computation efficiency under the frame of HOM-based beam position diagnostics. Equipped with comprehensive knowledge of HOMs in third harmonic cavities and potent analysis methods for position extraction, the dependencies of various dipole modes on transverse beam offsets in the cavity/module are studied in full swing in Chap. 5. Three modal options promising for beam position diagnostics are focused on: localized dipole beam-pipe modes at approximately 4 GHz, multi-cavity modes in the first two dipole bands within 4.5–5.5 GHz and trapped cavity modes in the fifth dipole band at approximately 9 GHz. A set of downconverter test electronics are subsequently designed and built in order to investigate position resolutions of these options. In Chap. 6, the principle of the test electronics and accuracies when determining both local position in the cavity and global position over the module are described along with an estimation of sources contributing to the resolution. This study constitutes a firm design of the HOMBPM electronics for third harmonic cavities at FLASH to be fixed. Chapter 7 gives a summary of the studies presented in this thesis and paves the way ahead of the HOMBPM system for third harmonic cavities.

1.1 Free-Electron Laser in Hamburg

Free-Electron Laser in Hamburg (FLASH) [6, 10] is a free-electron laser (FEL) facility at DESY in Hamburg, Germany. It is driven by a superconducting linear accelerator (linac). The production of FEL radiation is based on self-amplified spontaneous emission (SASE). FLASH provides trains of coherent very short (10–50 fs) laser pulses with an unprecedented brilliance. The photon wavelength range is tunable between 4 and 45 nm, covering the vacuum ultraviolet (VUV) to soft x-rays. FLASH

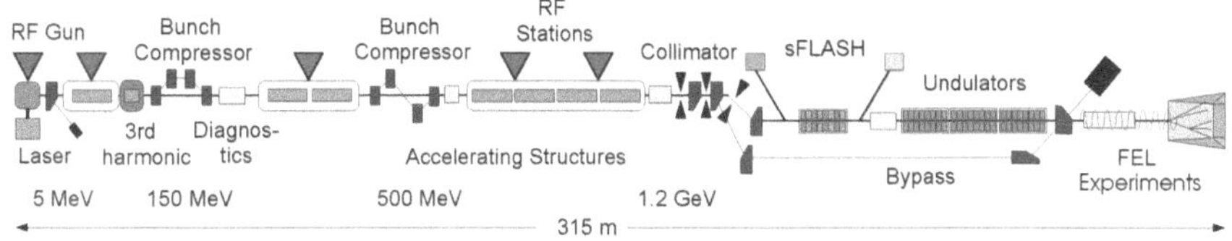

Fig. 1.1 Schematic of the FLASH facility. Figure courtesy of S. Schreiber

is a user facility for photon science and a test facility for various accelerator physics studies. It is also an important pilot facility for future projects based on superconducting accelerator technologies, like the European X-Ray Free-Electron Laser (European XFEL) [11] and the International Linear Collider (ILC) [12].

Figure 1.1 shows a schematic layout of the FLASH facility. A laser-driven normal conducting RF-gun produces electron bunch trains. Seven TESLA type superconducting accelerating modules [8] then accelerate the electron beam to maximum energy of 1.25 GeV. Eight 9-cell standing wave niobium cavities [7] with a fundamental frequency of 1.3 GHz are mounted into each of the 12 m-long cryo-modules. The cavities are bath-cooled by super-fluid Helium to 2 K. The first accelerating module boosts the initial 5 MeV beam energy to approximately 180 MeV. Then the third harmonic cavity module [13], containing four 9-cell 3.9 GHz cavities [9], linearizes the energy spread along the bunch and consequently reduces the beam energy to approximately 150 MeV [14]. The beam is subsequently compressed longitudinally by the first magnetic chicane followed by an acceleration with two modules boosting its energy to 500 MeV. Then the second magnetic chicane performs a further longitudinal bunch compression. Thereafter the last four modules accelerate the beam to the maximum energy of 1.25 GeV. The sFLASH located downstream of the collimator is an experiment aimed to study the high-gain-FEL amplification of a laser seed from a high harmonic generation (HHG) source [15]. Finally the SASE FEL radiation is produced by the electron beam traversing six 4.5 m long fixed gap undulators consisting of permanent magnets, and is delivered to the user FEL experiments.

The time structure of the FLASH beam is shown in Fig. 1.2. The spacing between bunches in a macropulse is variable between 1 MHz and 40 kHz. The number of bunches in a macropulse can be varied from 1 to 800 (with 1 MHz bunch spacing). The maximum length of a macropulse is 800 μs. The macropulse repetition rate is normally fixed to 10 Hz. Electron bunch charges between 0.1 and 1 nC are typically used in FEL operation, although charges up to 3 nC are possible as well.

1.2 Third Harmonic Cavities

The essential advantage of FEL radiation is its high intensity because a large number of electrons radiate coherently. This is realized by the process of microbunching, a modulation of the longitudinal velocity along the electron bunch which eventually

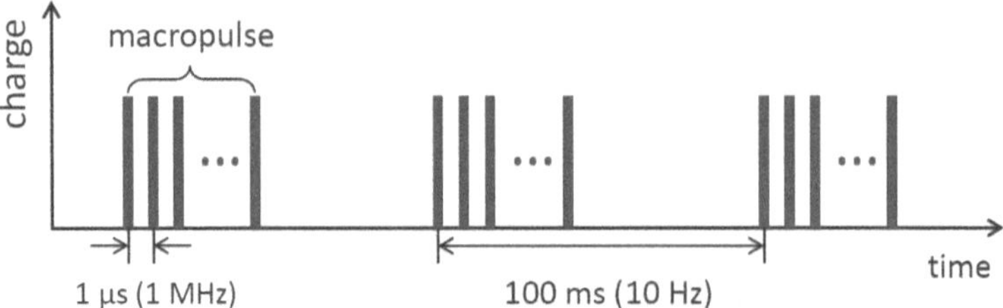

Fig. 1.2 The time structure of the electron beam at FLASH. Up to 800 bunches can be filled in a macropulse at a minimum spacing of 1 μs. The time between successive macropulses corresponds to the 10 Hz repetition rate of the RF system

packs the electrons into slices that are shorter than the photon wavelength. Inside an undulator, the FEL power exhibits an exponential growth after a certain distance, where the microbunching process is evident. The exponential regime ends when the microbuches are fully developed, and the FEL power saturates. This distance L_g where a FEL power grows by a factor $e = 2.718$, namely gain length, depends on the peak current I_{peak} and emittance ε of the electron beam in the form [16]

$$L_g = C \cdot \left(\frac{\gamma\varepsilon\beta}{I_{peak}}\right)^{\frac{1}{3}}, \quad (1.1)$$

where γ is the Lorentz factor, β is the beta function of the electron beam and C is a parameter determined by the undulator. From Eq. 1.1, high peak current and low emittance are desired in obtaining a short gain length and consequently a short undulator. The FEL wavelength λ_l is inversely proportional to the electron beam energy as [16]

$$\lambda_l \sim \frac{1}{\gamma^2}. \quad (1.2)$$

Therefore, a shorter FEL wavelength requires higher electron beam energy, which leads to an increasing gain length (see Eq. 1.1) and a longer undulator. Thus a high peak current and a low emittance of the electron beam is crucial for a high-gain FEL at short wavelength using a reasonable length of undulator. At FLASH the gain length is approximately 1 m, the I_{peak} is several kA, the beam size is less than 100 μm and the undulator is 27 m long [17].

We have addressed the importance of high peak currents (~kA) in high-gain X-ray FELs. However, it is impossible to produce ultra-short high-charge electron bunches with low transverse emittance directly by a RF gun due to strong space charge force at low energy (scales as $1/\gamma^2$). In order to preserve the required high quality emittance (~0.5 mm·mrad), long bunches of several millimeter with a peak current of 50 A are generated by the gun, quickly accelerated to higher energy and then compressed longitudinally by two orders of magnitude to reach the peak current of several kA.

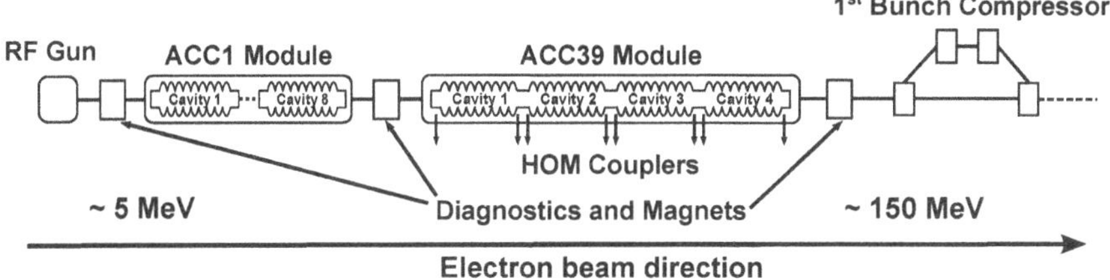

Fig. 1.3 Schematic of the FLASH injector section (not to scale). Reproduced with permission from Fig. 1.1 of [18]

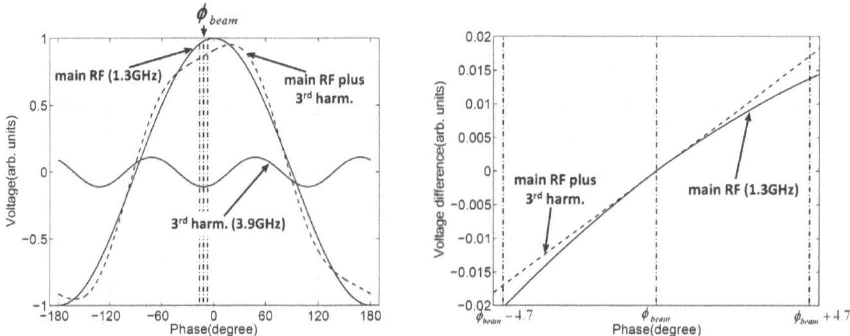

Fig. 1.4 Linearization of the RF electric field curvature with a third harmonic field [20]

Figure 1.3 shows schematically the injector section of FLASH, which is relevant to this study. The electron beam generated by a photoelectric gun is accelerated off-crest by eight superconducting TESLA 1.3 GHz cavities in cryo-module ACC1 and compressed longitudinally by the first magnetic chicane. Due to the length of the electron bunch in the millimeter range before the first bunch compressor, and the sinusoidal 1.3 GHz RF field, a curvature in the energy-phase plane develops, which leads to a long tail of the electron bunch and the reduction of peak current in the bunch compression process. To linearize the energy spread, harmonics of the fundamental accelerating frequency (1.3 GHz) of the main linac are added. At FLASH, third harmonic superconducting cavities operating at 3.9 GHz are used [14]. The overall field profile with and without the additional third harmonic is illustrated in Fig. 1.4. The measured phase space linearization is shown in Fig. 1.5. It can be clearly seen that the non-linearity in longitudinal phase space is removed by turning on the 3.9 GHz cavities. The details of this measurement can be found in Ref. [19].

Four such cavities are placed in the ACC39 module following ACC1 (see Fig. 1.6), which was designed and built by Fermilab in collaboration with DESY [21, 13]. Figure 1.7 shows a photo of ACC39 in the FLASH beam line along with some essential parameters for this cryo-module.

The third harmonic cavity inherits a similar design of the TESLA 1.3 GHz cavity. A schematic of the third harmonic cavity is illustrated in Fig. 1.8a along with the main

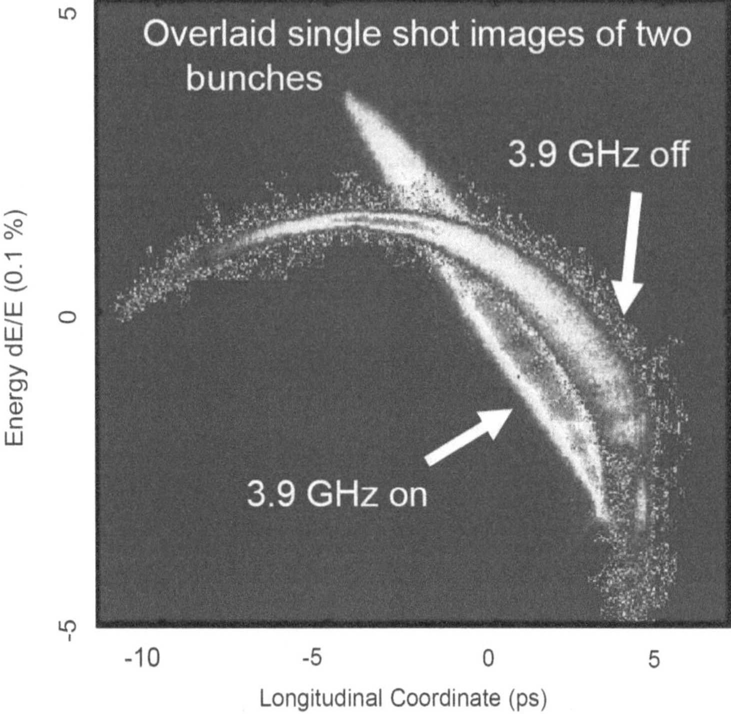

Fig. 1.5 The first linearization of the longitudinal phase space by applying 15 MV 3.9 GHz voltage. Figure courtesy of H. Edwards [19]

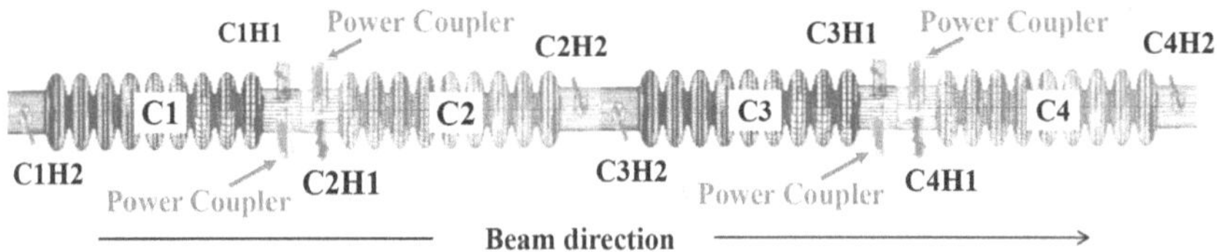

Fig. 1.6 Schematic of the four cavities within ACC39 module. The power couplers are placed downstream for C1 and C3, and upstream for C2 and C4. The HOM couplers located on the same side of the power couplers are named H1, while the other ones H2. Reprinted with permission from [22]. Copyright 2012, American Institute of Physics

dimensions. It has one power coupler and one pick-up probe installed on the beam pipes connecting the end-cells. It is also equipped with two HOM couplers installed on either side of the connecting beam pipes with different rotations as shown in Fig. 1.8b. The energy of HOMs are extracted by using these couplers and guided by long cables to the HOM absorbers on the patch panel. Consequently their deleterious

Parameters of ACC39	
Number of cavities	4
Active cavity length	0.346 m
Designed gradient	14 MV/m
Operating frequency	3.9 GHz

Fig. 1.7 ACC39 module in FLASH. The module on the *right* is ACC1. The direction of travel of the multi-bunch beam, is from *right to left*. Figure courtesy of DESY

effects on the beam are minimized. There are two distinct HOM coupler designs in order to cure the multipacting issues observed in the original design [23]. The HOM couplers with one post (Fig. 1.9a) are mounted on the first cavity (C1) and the third cavity (C3), while the two post ones (Fig. 1.9b) on the second cavity (C2) and the fourth cavity (C4).

By design, the 3.9 GHz cavity has two particularly challenging features in terms of wakefields, compared to the 1.3 GHz cavity. First, wakefields in the 3.9 GHz cavity are larger than in the 1.3 GHz cavity as the iris radius is much smaller: 15 mm in the 3.9 GHz cavity compared to 35 mm in the 1.3 GHz cavity (see Fig. 1.10). From scaling considerations, it can be shown that wakefields in the cavity grow as [25]

$$W_{\parallel} \sim \omega^2, \tag{1.3a}$$

$$W_{\perp} \sim \omega^3, \tag{1.3b}$$

where $W_{\parallel}$ and $W_{\perp}$ are the longitudinal and transverse wakefield respectively, and $\omega/2\pi$ is the frequency. Second, the HOM spectrum is significantly more complex than that of the 1.3 GHz cavity. The main reason for this is that unlike the TESLA cavity case, the majority of the modes are above the cutoff frequencies of the interconnecting beam pipes. This allows most of the modes from each independent cavity to propagate through to adjacent cavities. In this case, most modes reach all eight HOM couplers. This facilitates the damping of HOMs to be distributed. The overall damping mechanism may be superior to that achieved in TESLA cavities, although detailed comparisons have not been made [9]. On the other hand, this gives rise to a complexity in the HOM spectrum of the third harmonic cavity as will be seen in Chap. 3.

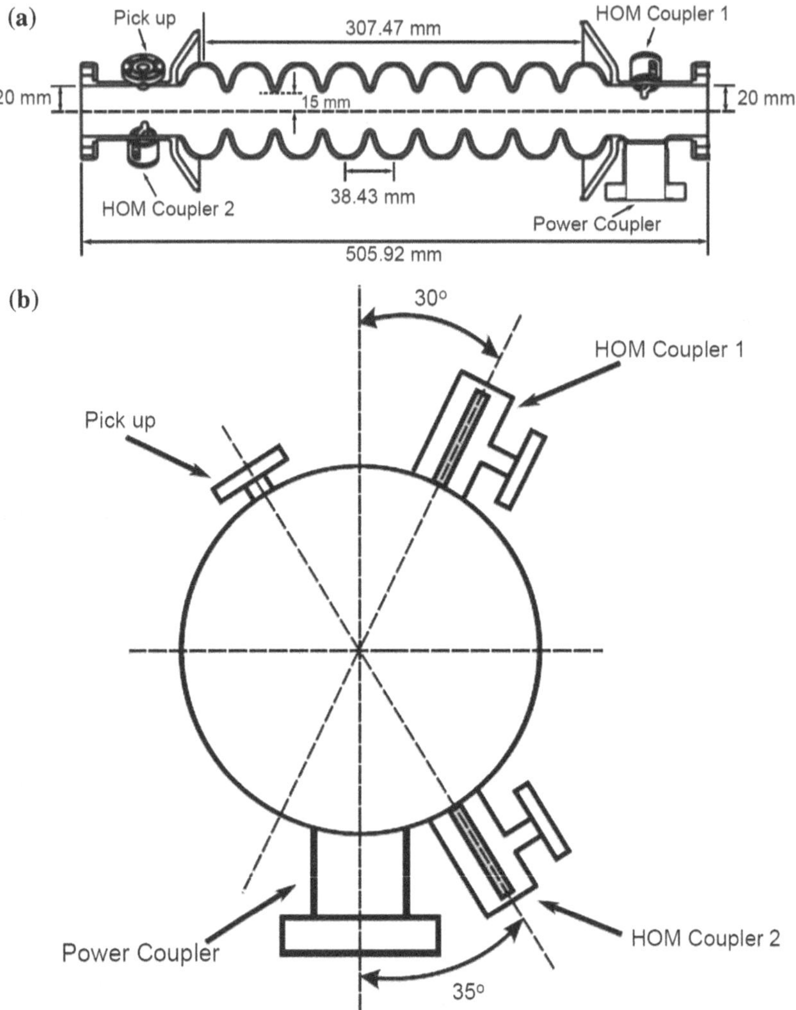

Fig. 1.8 Schematic of a third harmonic cavity with one power coupler, one pick up probe and two HOM couplers. Reproduced with permission from Fig. 1.2 of [18] **a** 3.9 GHz cavity **b** Coupler orientations

1.3 Wakefields

When a charged particle beam travels through an accelerating cavity, its generated electromagnetic field interacts with its surroundings. This scattered field is referred to as a wakefield.

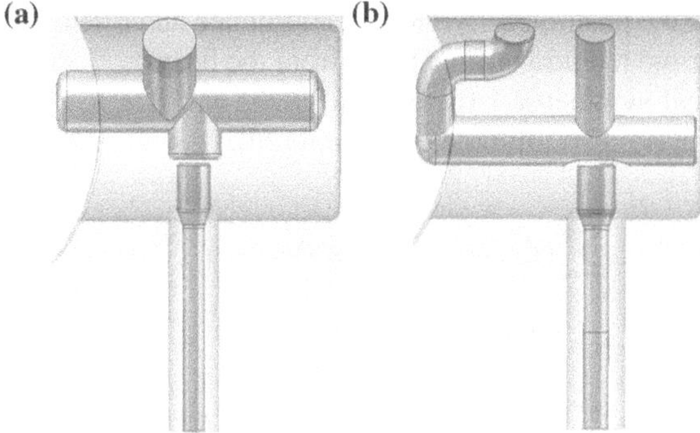

Fig. 1.9 **a** 1-post HOM coupler. **b** 2-post HOM coupler. 1-post and 2-post HOM coupler design made in CST Microwave Studio®[24]. The couplers with 1 post are: C1H1, C1H2, C3H1 and C3H2; The couplers with 2 posts are: C2H1, C2H2, C4H1 and C4H2. Figure courtesy of T. Flisgen

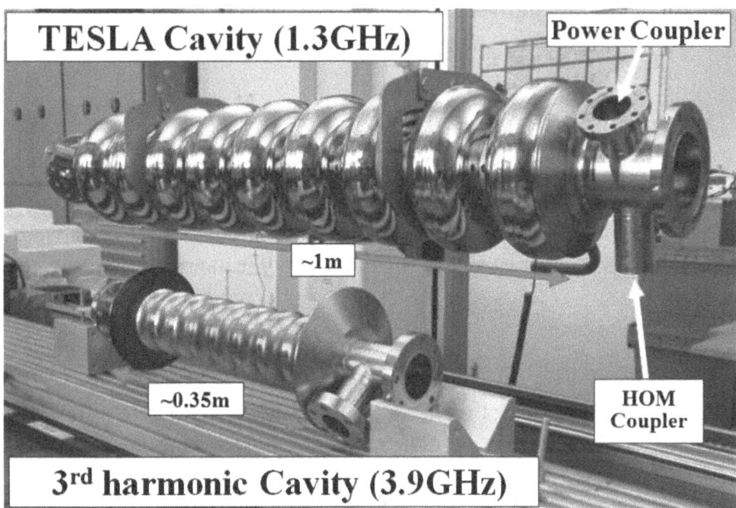

Fig. 1.10 A TESLA type cavity operating at 1.3 GHz (*top*) and the corresponding third harmonic 3.9 GHz cavity (*bottom*). Reproduced and courtesy of DESY

1.3.1 Electromagnetic Field in a Cavity

Modes in a cavity are resonant electromagnetic fields. For a mode with the frequency of $f = \omega/(2\pi)$, the electric and magnetic field can be written as

$$\mathbf{E}(r, \theta, z, t) = \mathbf{E}(r, \theta, z)e^{-i\omega t}, \tag{1.4}$$

$$\mathbf{B}(r, \theta, z, t) = \mathbf{B}(r, \theta, z)e^{-i\omega t}, \tag{1.5}$$

where (r, θ, z) is the position in cylindrical coordinate and t denotes the time which has elapsed after the radiation of the fields from the beam.

In a cylindrical symmetric structure, $\mathbf{E}(r, \theta, z)$ and $\mathbf{B}(r, \theta, z)$ can be expanded into a series of modes by multipole expansion [26],

$$\begin{aligned}\mathbf{E}(r,\theta,z) = \sum_{m=0}^{\infty}(&E_r^{(m)}(r,z)cos(m\theta)\mathbf{e_r} + E_\theta^{(m)}(r,z)sin(m\theta)\mathbf{e}_\theta \\ &+ E_z^{(m)}(r,z)cos(m\theta)\mathbf{e_z}), \end{aligned} \tag{1.6}$$

$$\begin{aligned}\mathbf{B}(r,\theta,z) = \sum_{m=0}^{\infty}(&B_r^{(m)}(r,z)sin(m\theta)\mathbf{e_r} + B_\theta^{(m)}(r,z)cos(m\theta)\mathbf{e}_\theta \\ &+ B_z^{(m)}(r,z)sin(m\theta)\mathbf{e_z}), \end{aligned} \tag{1.7}$$

where $m = 0, 1, 2, 3$ corresponds to monopole, dipole, quadrupole and sextupole modes, $\mathbf{e_r}$, $\mathbf{e}_\theta$ and $\mathbf{e_z}$ are unit vectors in r, θ and z direction respectively.

1.3.2 Definitions of RF Parameters

In this section, we will define several RF parameters which are commonly used to characterize the accelerating cavity and beam-cavity interactions. For a specific mode with index m and the frequency of ω, its quality factor in an accelerating cavity, Q, is defined as

$$Q^{(m)} = \frac{\omega U^{(m)}}{P}, \tag{1.8}$$

where P is the average power loss in the walls of the cavity due to a finite electrical resistance, $U^{(m)}$ is the total energy of the mode (m) stored in the cavity and is given as

$$U^{(m)} = \frac{\epsilon_0}{2}\int d^3r|\mathbf{E^{(m)}}|^2, \tag{1.9}$$

where ϵ_0 is the vacuum permittivity. The electric field $\mathbf{E^{(m)}}$ is defined as

$$\begin{aligned}\mathbf{E^{(m)}} = &E_r^{(m)}(r,z)cos(m\theta)\mathbf{e_r} + E_\theta^{(m)}(r,z)sin(m\theta)\mathbf{e}_\theta \\ &+ E_z^{(m)}(r,z)cos(m\theta)\mathbf{e_z}, \end{aligned} \tag{1.10}$$

where $E_r^{(m)}$, $E_\theta^{(m)}$ and $E_z^{(m)}$ have been defined in Eq. 1.6.

The longitudinal voltage of the mode (m), $V_\parallel^{(m)}$, at radius r is obtained by a line integral of the longitudinal electric field along a dedicated path in z direction as

$$V_{\|}^{(m)} = \int_0^L dz \cdot E_z^{(m)}(r, z) e^{-i\omega z/c}, \tag{1.11}$$

where L is the length of the longitudinal path, and the particle moves with the velocity c.

For a given power loss P, the shunt impedance of the mode measures the efficiency of producing a longitudinal voltage $V_{\|}^{(m)}$ in the cavity. It is defined as

$$R^{(m)} = \frac{|V_{\|}^{(m)}|^2}{P}. \tag{1.12}$$

The ratio of shunt impedance to quality factor, $(R/Q)^{(m)}$, is a parameter to characterize the interaction between the beam and a cavity mode. It is often normalized with the beam offset r and is calculated as

$$\left(\frac{R}{Q}\right)^{(m)} = \frac{1}{r^{2m}} \frac{|V_{\|}^{(m)}|^2}{2\omega U^{(m)}}. \tag{1.13}$$

A factor of 1/2 is added by convention. The $(R/Q)^{(m)}$ of a monopole mode ($m = 0$) has the unit of $[\Omega]$, while a dipole mode ($m = 1$) has $[\Omega \cdot \text{cm}^{-2}]$.

1.3.3 Definition of Wakefields

As shown in Fig. 1.11, an excitation particle of charge q' is moving in the z direction in an accelerating structure, followed by a test particle with charge q at distance s. Both particles are point charge and ultra-relativistic (velocity $\approx c$). The electromagnetic field generated by the excitation particle q' acts on the test particle q in the form of Lorentz force as

$$\mathbf{F} = q(\mathbf{E} + c\mathbf{e_z} \times \mathbf{B}). \tag{1.14}$$

The wake potential is defined as the total voltage experienced by a test charge following at a distance s behind the excitation particle along the length L of the structure, normalized by the charge of the excitation particle,

$$\begin{aligned} \mathbf{W}(s, r', \theta', r, \theta) &= -\frac{1}{qq'} \int_0^L dz \cdot \mathbf{F} \\ &= -\frac{1}{q'} \int_0^L dz \cdot (\mathbf{E} + c\mathbf{e_z} \times \mathbf{B}). \end{aligned} \tag{1.15}$$

Since a signal cannot travel faster than the speed of light, $\mathbf{W}(s) = 0$ for $s < 0$. The distance s is positive in the direction opposite to the motion of the excitation particle q'. The longitudinal wake potential $\mathbf{W}_{\|}$ is defined as

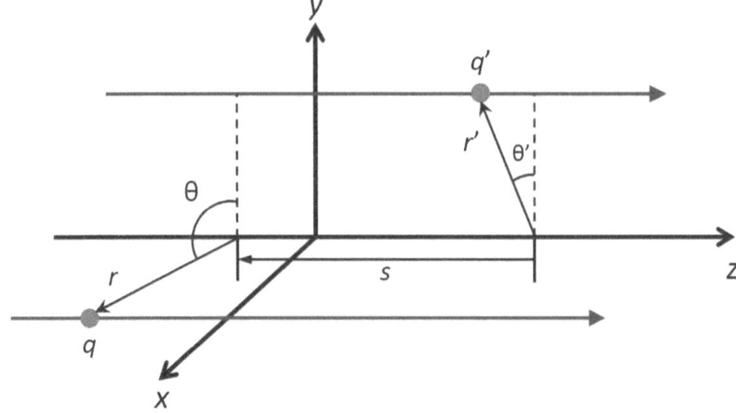

Fig. 1.11 A point charged particle q' traversing a cavity with an offset r' and azimuthal angle θ' followed by a test charged particle q with offset r and azimuthal angle θ. The Figure was inspired by [27]

$$\mathbf{W}_{\parallel}(s, r', \theta', r, \theta) = -\frac{1}{q'} \int_0^L dz \cdot E_z(z, (z+s)/c)\mathbf{e_z}, \tag{1.16}$$

where $E_z(z, (z+s)/c)$ is the z component of the electric field $\mathbf{E}(r, \theta, z, t)$ on the path of the test particle q. Similarly the transverse wake potential $\mathbf{W}_{\perp}$ is defined as the transverse momentum experienced by the test particle normalized by the charge q',

$$\mathbf{W}_{\perp}(s, r', \theta', r, \theta) = \frac{1}{q'} \int_0^L dz \cdot [\mathbf{E}_{\perp} + (c\mathbf{e_z} \times \mathbf{B})_{\perp}]_{t=(z+s)/c}. \tag{1.17}$$

Note that $\mathbf{W}_{\perp}$ is a vector with both r and θ components. The longitudinal and transverse wake potentials are connected by the Panofsky-Wenzel theorem [28],

$$\frac{\partial}{\partial s}\mathbf{W}_{\perp}(s, r', \theta', r, \theta) = -\nabla_{\perp} W_{\parallel}(s, r', \theta', r, \theta), \tag{1.18}$$

where $\nabla_{\perp}$ is applied on the transverse coordinates of the test particle q: r and θ. The transverse wake potential can be obtained by the integration of the transverse gradient of the longitudinal wake potential. This theorem applies also to structures that are not cylindrically symmetric.

In a cylindrical symmetric structure, the wake potential can be decomposed by a multipole expansion as [26]

$$\mathbf{W}_{\parallel}(s, r', \theta', r, \theta) = \sum_{m=0}^{\infty} (r')^m r^m W_{\parallel}^{(m)}(s) cos\left(m(\theta - \theta')\right) \mathbf{e_z}, \tag{1.19}$$

$$\mathbf{W}_{\perp}(s, r', \theta', r, \theta) = \sum_{m=1}^{\infty} m(r')^m r^{m-1} W_{\perp}^{(m)}(s) \cdot \left(\mathbf{e_r} cos\left(m(\theta - \theta')\right) - \mathbf{e}_{\theta} sin\left(m(\theta - \theta')\right)\right), \quad (1.20)$$

where $m = 0, 1, 2$ corresponds to monopole, dipole and quadrupole modes, $W_{\parallel}^{(m)}(s)$ and $W_{\perp}^{(m)}(s)$ are the longitudinal and transverse m-pole wake potentials respectively. Equation 1.20 is obtained by substituting Eq. 1.19 into Eq. 1.18. The longitudinal and transverse m-pole wake potential are connected by

$$W_{\perp}^{(m)}(s) = -\int_{-\infty}^{s} ds' \cdot W_{\parallel}^{(m)}(s'), m > 0. \quad (1.21)$$

Since particles normally remain near axis, it is often sufficient to consider only the leading terms of the series in Eqs. 1.19 and 1.20, neglecting contributions from higher multipole components.

$$W_{\parallel}(s, r', \theta', r, \theta) = W_{\parallel}^{(0)}(s) + r'r W_{\parallel}^{(1)}(s) cos(\theta - \theta') + \cdots, \quad (1.22)$$

$$\mathbf{W}_{\perp}(s, r', \theta', r, \theta) = r' W_{\perp}^{(1)}(s)(\mathbf{e_r} cos(\theta - \theta')) - \mathbf{e}_{\theta} sin(\theta - \theta')) + \cdots. \quad (1.23)$$

There is no transverse monopole ($m = 0$) wake potential. The dipole wake potential does not depend on the transverse position of the test particle q. The kick on the test particle is linear to the offset of the excitation particle q'. The longitudinal wake potential does not give a transverse kick.

It can be shown that the m-pole wake potential can be written as a sum over all the m-pole modes [29]. The longitudinal m-pole wake potential is

$$W_{\parallel}^{(m)}(s) = -\sum_{n} \omega_{mn} \left(\frac{R}{Q}\right)^{(mn)} cos(\omega_n s/c) H(s), \quad (1.24)$$

where $H(s)$ is the Heaviside step function [30], $(R/Q)^{(mn)}$ and ω_{mn} are the R over Q and frequencies of the nth m-pole modes respectively. In an infinite long periodic structure the modes come in frequency passbands, though a charged particle travelling through the structure with a limited number of cells will only excite a discrete set of these modes, grouped in bands. The transverse m-pole wake potential according to Eqs. 1.24 and 1.21 is

$$W_{\perp}^{(m)}(s) = c \sum_{n} \left(\frac{R}{Q}\right)^{(mn)} sin(\omega_{mn} s/c) H(s), m > 0. \quad (1.25)$$

By substituting Eqs. 1.24 and 1.25 into Eqs. 1.22 and 1.23, and taking the leading term, the longitudinal wake potential is then dominated by monopole modes ($m = 0$), while dipole modes ($m = 1$) dominate the transverse wake potential. They are written as

$$\mathbf{W}_{\parallel} \simeq -\sum_{n} \omega_{mn} \left(\frac{R}{Q}\right)^{(0n)} cos(\omega_{0n} s/c) H(s) \mathbf{e_z}, \quad (1.26)$$

$$\mathbf{W}_{\perp} \simeq r' c(\mathbf{e_r} cos(\theta - \theta') - \mathbf{e}_{\theta} sin(\theta - \theta')) \sum_{n} \left(\frac{R}{Q}\right)^{(1n)} sin(\omega_{1n} s/c) H(s). \quad (1.27)$$

The longitudinal wake potential is approximately independent of the transverse position of both the excitation particle and the test particle. The transverse wake potential depends on the excitation particle as the first power of its offset r', and independent of the test particle's transverse position r. The amplitude of a specific dipole mode signal A_{dipole} is proportional to its $(R/Q)^{(1n)}$, the beam offset r' and the excitation particle's charge q',

$$A_{dipole} = q' \cdot |W_{\perp}| \sim q' \cdot r' \cdot \left(\frac{R}{Q}\right)^{(1n)}. \quad (1.28)$$

The phase of the mode is determined by the sign of r'. This enables the beam position within the cavity to be remotely determined by monitoring the beam-excited dipole modes [31–33].

In Chap. 2, the HOMs of the third harmonic cavities are studied by simulations. The modal band structure of the cavity is characterized. This serves as the first step in the study of beam position diagnostics with HOMs.

References

1. B.W. Zotter and S.A. Kheifets, *Impedances and Wakes in High Energy Particle Accelerators*, 1st edn. (World Scientific, Singapore, 1998)
2. K. Yokoya, Cumulative beam breakup in large scale linacs. DESY Report: 86–084, 1986
3. J.S. Sekutowicz, *HOM Damping and Power Extraction from Superconducting Cavities*, in *Proceedings of LINAC 2006*, (*Knoxville* (Tennessee, 2006), pp. 506–510
4. R.M. Jones, Wake field suppression in high gradient linacs for lepton linear colliders. Phys. Rev. ST Accel. Beams **12**, 104801 (2009)
5. T. Weiland, R. Wanzenberg, Wake fields and impedances. Lect. Notes Phys. **400**, 39–79 (1992)
6. W. Ackermann et al., Operation of a free-electron laser from the extreme ultraviolet to the water window. Nature Photonics **1**, 336–342 (2007)
7. J. Sekutowicz, *Multi-cell Superconducting Structures for High Energy e+e- Colliders and Free Electron Laser Linacs*, 1st edn. (Warsaw University of Technology Publishing House, Warsaw, 2008)
8. TESLA Technical Design Report. DESY Report: DESY 01–011, 2001
9. J. Sekutowicz, R. Wanzenberg, W.F.O. Mueller and T. Weiland, A Design of a 3rd Harmonic Cavity for the TTF 2 Photoinjector. TESLA-FEL Report: TESLA-FEL 2002–2005, 2002
10. K. Honkavaara et al., *Status of FLASH*, in *Proceedings of IPAC2012*, (Louisiana, 2012), pp. 1715–1717

11. The European X-Ray Free-Electron Laser. Technical design report: DESY 2006–097, 2007
12. International Linear Collider: A Technical Progress Report, Interim Report, 2011
13. E.R. Harms et al., *Third Harmonic System at Fermilab/FLASH*, in *Proceedings of SRF2009*, (Berlin, 2009), pp. 11–16
14. K. Floettmann, T. Limberg and Ph. Piot, Generation of Ultrashort Electron Bunches by Cancellation of Nonlinear Distortions in the Longitudinal Phase Space. TESLA-FEL Report: TESLA-FEL 2001–06, 2001
15. V. Miltchev et al., *sFLASH—Present Status and Commissioning Results*, in *Proceedings of IPAC2011*, (San Sebastian, 2011), pp. 923–927
16. P. Schmueser, *Free-Electron Lasers*. Lecture Note: CERN Accelerator School (Zeuthen, Germany, 2003)
17. The TESLA Test Facility FEL team, SASE FEL at the TESLA Facility, Phase2, TESLA-FEL Report: TESLA-FEL 2002–01, 2002
18. P. Zhang, N. Baboi, R.M. Jones, N. Eddy, Resolution study of higher-order-mode-based beam position diagnostics using custom-built electronics in strongly coupled 3.9 GHz multi-cavity accelerating module. J. Instrum. **7**, P11016 (2012)
19. H. Edwards et al., *3.9 GHz Cavity Module for Linear Bunch Compression at FLASH*, in *Proceedings of LINAC2010*, (Tsukuba, 2010), pp. 41–45
20. E. Vogel et al., *Considerations on the third harmonic rf of the European XFEL*, in *Proceedings of SRF2007*, (Beijing, 2007), pp. 481–485
21. E. Vogel et al., *Test and Commissioning of the Third Harmonic RF System for FLASH*, in *Proceedings of IPAC'10*, (Kyoto, Japan, 2010), pp. 4281–4283
22. P. Zhang, N. Baboi, R.M. Jones, I.R.R. Shinton, T. Flisgen, H.W. Glock, A study of beam position diagnostics using beam-excited dipole modes in third harmonic superconducting accelerating cavities at a free-electron laser. Rev. Sci. Instrum. **83**, 085117 (2012)
23. T. Khabibouline et al., *New HOM Coupler Design for 3.9 GHz Superconducting Cavities at FNAL*, in *Proceedings of PAC07* (Albuquerque, New Mexico, 2007), pp. 2259–2261
24. C.S.T. Microwave Studio®. Ver. 2011, CST AG (Darmstadt, Germany, 2011)
25. K.L.F. Bane, Wake Field Effects in a Linear Collider. SLAC Note: SLAC-PUB-4169, 1986
26. A. Chao, *Physics of Collective Beam Instabilities in High Energy Accelerators*, 1st edn. (Wiley, New York, 1993)
27. R.M. Jones, *HOM Mitigation*. Lecture Note: CERN Accelerator School (Ebeltoft, Denmark, 2010)
28. W.K.H. Panofsky, W.A. Wenzel, Some considerations concerning the transverse deflection of charged particles in radio-frequency fields. Rev. Sci. Instrum. **27**, 967 (1956)
29. E.U. Condon, Forced oscillations in cavity resonators. J. Appl. Phys. **12**, 129 (1941)
30. R. Bracewell, *The Fourier Transform and its Applications*, 3rd edn. (McGraw-Hill, New York, 2000), p. 61
31. G. Devanz et al., *HOM Beam Coupling Measurements at the TESLA Test Facility (TTF)*, in *Proceedings of EPAC2002*, (Paris, 2002), pp. 230–232
32. N. Baboi et al., *Preliminary Study on HOM-Based Beam Alignment in the TESLA Test Facility*, in *Proceedings of LINAC 2004*, (Lübeck, France), pp. 117–119, 2004
33. S. Molloy et al., High precision superconducting cavity diagnostics with higher order mode measurements. Phys. Rev. ST Accel. Beams **9**, 112801 (2006)

Chapter 2
Electromagnetic Eigenmode Simulations of the Third Harmonic Cavity

In order to study the nature of the HOMs in third harmonic cavities present in ACC39, eigenmode simulations have been conducted. This is essential and the first step of searching suitable modes for beam position diagnostics. Eigenmodes of an ideal third harmonic 9-cell cavity are used to study the modal structure of each band and to interpret the HOM spectra measurements described in Chap. 3.

In this chapter, the third harmonic cavity is first treated as a periodic structure with an infinite number of repetitions of the mid-cell. The dispersion curves of monopole, dipole, quadrupole and sextupole passbands are described in Sect. 2.1. The beam pipes connecting cavities modeled as circular waveguides are described in Sect. 2.2. The eigenmodes obtained for an ideal third harmonic cavity without couplers are presented in Sect. 2.3.

2.1 The Third Harmonic Cavity as a Periodic Structure

A sketch of the cell geometry of the third harmonic cavity is given in Fig. 2.1. The cell is rotationally symmetric around the z axis. The iris and the equator both have an elliptical shape. The mid-cell has a different shape compared to the end-cell, and parameters are listed in Table 2.1. The iris of the end-cup is larger than that of the mid-cup in order to accomplish a better coupling [1].

A mid-cell subjected to infinite periodic boundary conditions was simulated. Fig. 2.2a shows a mid-cell modeled in CST Microwave Studio® [2]. A hexahedral mesh was used in the calculation of the electromagnetic field as shown in Fig. 2.2b. The mesh lines were chosen such that the iris radius and the equator radius were exactly matched by mesh lines. Symmetry planes were applied on the structure to save simulation time so that only a quarter of the structure was used (see the region with mesh lines in Fig. 2.2b). Approximately 130,000 mesh cells for a quarter of the structure and a maximum mesh step of 0.85 m were set. Electric (EE) boundary conditions were used on the surface of the mid-cell, while periodic boundary conditions were set on both ends of the cell. A solver accuracy of 10^{-6} in terms of the eigensystem's relative residual was used.

P. Zhang, *Beam Diagnostics in Superconducting Accelerating Cavities*, Springer Theses, DOI: 10.1007/978-3-319-00759-5_2,

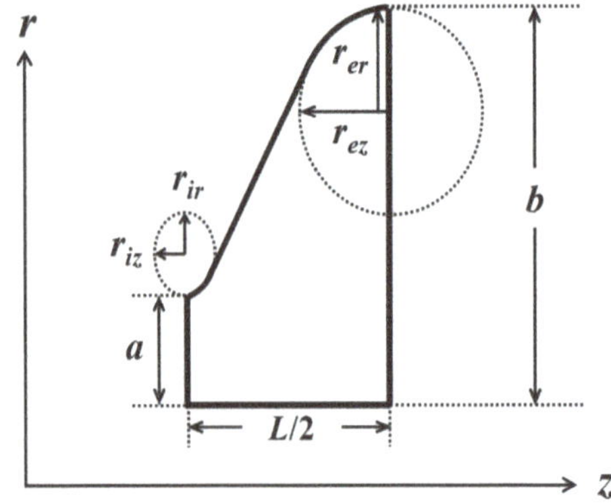

Fig. 2.1 Parameterization of the cell geometry [1]. The *solid curve* represents the cell wall

Table 2.1 Parameters of the cell geometry of the third harmonic cavity [1]

		mid-cup	end-cup
Iris radius, $\boldsymbol{a}$	mm	15.0	20.0
Equator radius, $\boldsymbol{b}$	mm	35.787	35.787
Half cell length, $\boldsymbol{L}/2$	mm	19.2167	19.2167
Equator horizontal axis, $\boldsymbol{r_{ez}}$	mm	13.6	14.4
Equator vertical axis, $\boldsymbol{r_{er}}$	mm	15.0	15.0
Iris horizontal axis, $\boldsymbol{r_{iz}}$	mm	4.5	4.5
Iris vertical axis, $\boldsymbol{r_{ir}}$	mm	6.0	6.0

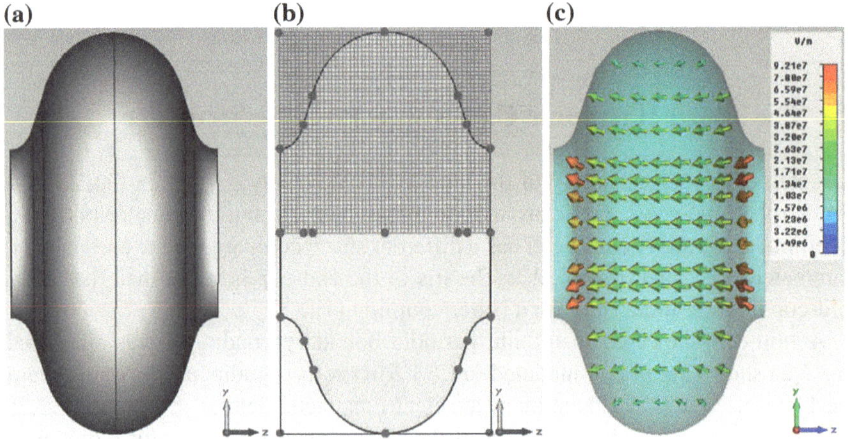

Fig. 2.2 **a, b** The mid-cell of the third harmonic cavity as modeled in CST Microwave Studio®. **c** The electric field of the accelerating mode with a phase advance of 180 degrees per cell (π mode)

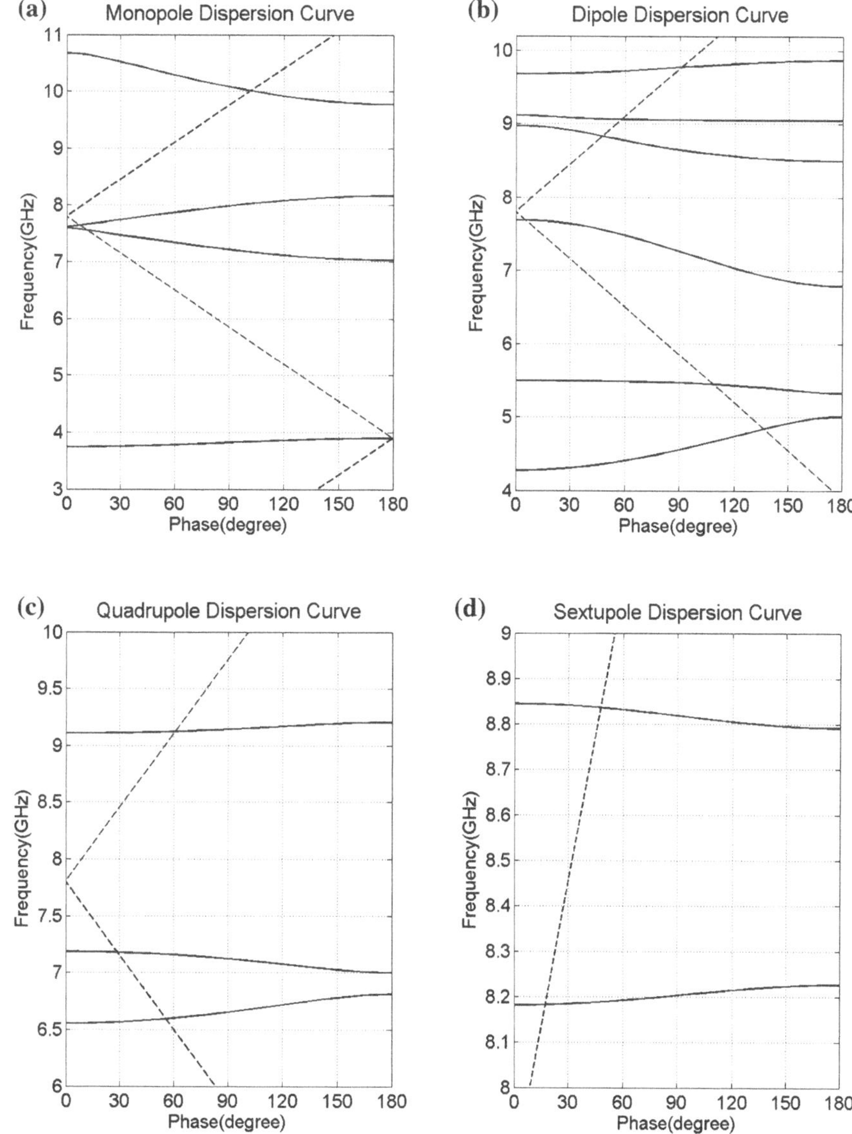

Fig. 2.3 The band structure (*solid curve*) of a 3.9 GHz cavity mid-cell. The light line is *dashed*

The modes of an infinitely long periodic chain of cavities can be obtained from single cell calculations using Floquet periodic boundary conditions:

$$\mathbf{E}(r, z + L) = \mathbf{E}(r, z)e^{i\phi}, \tag{2.1}$$

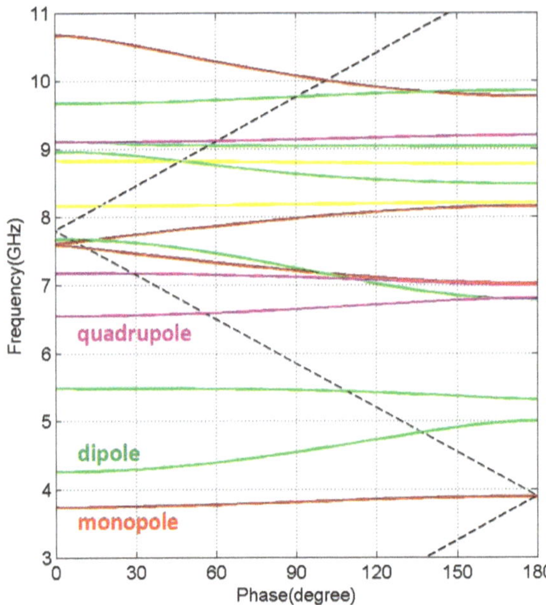

Fig. 2.4 Dispersion curve for monopole (*red*), dipole (*green*), quadrupole (*magenta*) and sextupole (*yellow*) modes. The light line is *dashed*

Table 2.2 Values of p_{mn} and p'_{mn} [3, 4]

	p_{mn} (TM modes)				p'_{mn} (TE modes)		
	m	n = 1	n = 2	n = 3	n = 1	n = 2	n = 3
Monopole	0	**2.405**	5.520	8.654	3.832	7.016	10.174
Dipole	1	3.832	7.016	10.174	**1.841**	5.331	8.536
Quadrupole	2	5.136	8.417	11.620	3.054	6.706	9.970
Sextupole	3	6.380	9.761	13.015	4.201	8.015	11.346

Table 2.3 Cutoff frequencies for the lowest order TE and TM modes in a circular waveguide with a radius of 15 and 20 mm

	a = 15 mm (GHz)	**a** = 20 mm (GHz)
f_c (TE_{11})	5.86	4.39
f_c (TM_{01})	7.65	5.74

where L is the cell length and ϕ is the phase advance per cell. Figure 2.2c shows the electric field of a mode with a phase advance of 180 degrees per cell, which is the so-called π mode. The frequencies of several passbands are shown in the form of dispersion curves [3] in Fig. 2.3 for monopole, dipole, quadrupole and sextupole modes. The phase velocity is defined as [3]

$$v_{phase} = \frac{\omega}{k_z} = 2\pi L \frac{f}{\phi}, \tag{2.2}$$

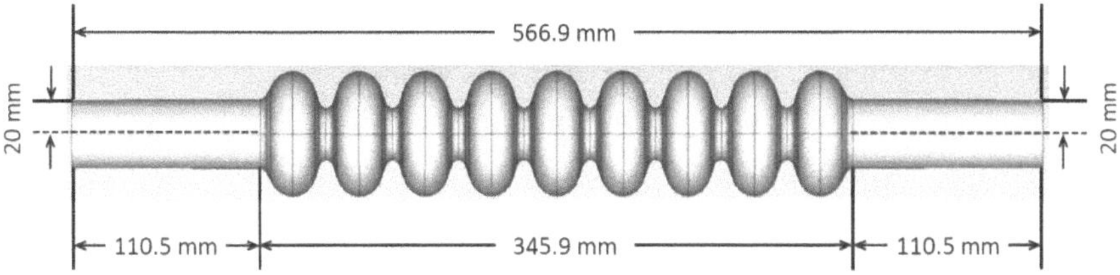

Fig. 2.5 CST Microwave Studio® generated geometry of the third harmonic cavity

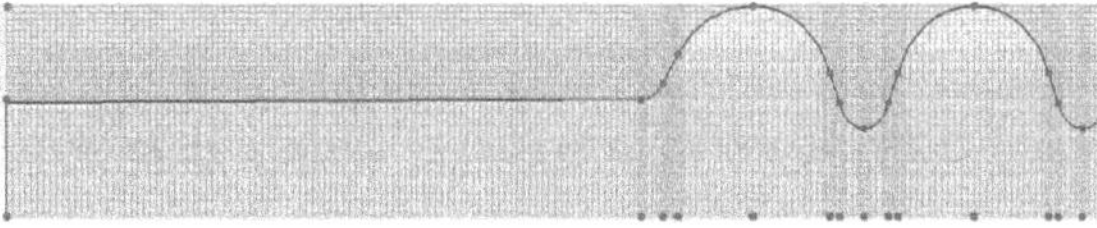

Fig. 2.6 A typical mesh used for selected cells in the third harmonic cavity

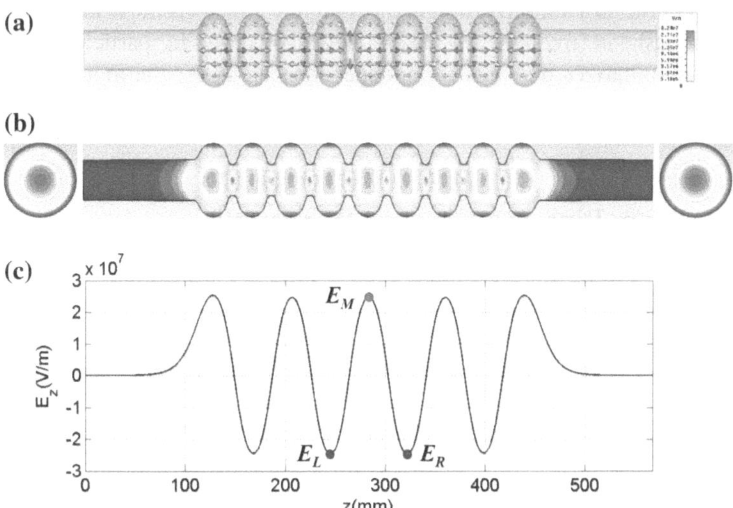

Fig. 2.7 **a** The electric field (*arrows*) of the accelerating mode (3.9 GHz) in the third harmonic cavity. **b** The electric field magnitude of the accelerating mode (frequency: 3.9008 GHz, R/Q: 373.113 Ω). Electric (EE) boundary conditions were used in the simulation. **c** Longitudinal electric field E_z of the accelerating mode (3.9 GHz, π mode) on the cavity axis. The *dots* are corresponding to the position marked in Fig. 2.8 for E_L, E_M and E_R respectively

where k_z is the longitudinal wave number, $\phi = k_z L$ is the phase advance per cell, which is used as a horizontal axis in the plots of the dispersion curves. The phase velocity can be obtained from the dispersion curves using Eq. 2.2.

A beam excites strongest those modes which are synchronous to the beam, i.e. with a phase velocity equal to the speed of the accelerated particles. For FLASH, this is the speed of light, $v_{phase} = c$. The so-called light line can therefore be drawn as the straight line:

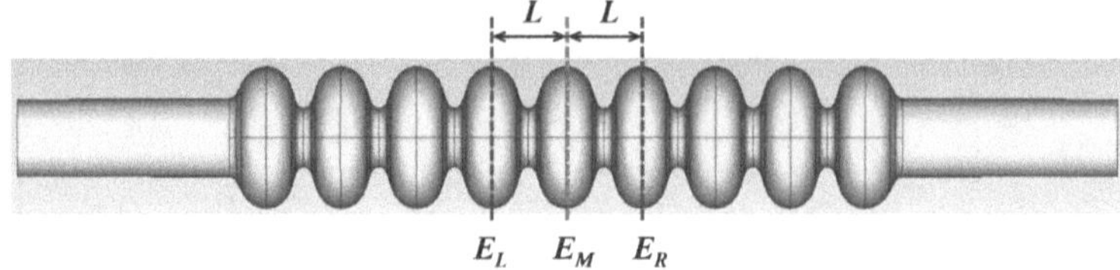

Fig. 2.8 Calculation of the phase advance per cell. E_L, E_M and E_R are longitudinal electric field at certain positions

$$f(\phi) = \frac{c}{2\pi L}\phi. \tag{2.3}$$

It is folded into the phase range from 0 to 180 degrees in the dispersion plots due to the periodicity of the structure. By design, the light line intersects the π mode of the first monopole passband (frequency $\approx$ 3.9 GHz), which is used for particle acceleration. Figure 2.4 summarizes the calculated dispersion curves for the monopole, dipole, quadrupole and sextupole bands.

2.2 The Beam Pipe as a Circular Waveguide

The third harmonic cavities are connected with beam pipes, whose radius is equal to the iris of the end-cup. To study the propagation of modes amongst cavities, the beam pipes are treated as infinitely long circular waveguides, which can be characterized analytically. Generally, the transverse electric (TE) and transverse magnetic (TM) modes can be distinguished from the characterization of the electric and magnetic fields [3]. The cutoff frequencies of TE and TM modes for a circular waveguide are [4]:

$$f_c(\mathrm{TM_{mn}}) = c\frac{p_{mn}}{2\pi a}, \tag{2.4a}$$

$$f_c(\mathrm{TE_{mn}}) = c\frac{p'_{mn}}{2\pi a}, \tag{2.4b}$$

where $m = 0, 1, 2, 3$ corresponds to monopole, dipole, quadrupole and sextupole modes, p_{mn} is the nth root of the mth Bessel function J_m, p'_{mn} is the nth root of the derivative of the mth Bessel function J'_m, a is the radius of the waveguide. The first TE mode to propagate is the mode with the smallest p'_{mn}, which from Table 2.2 is seen to be $\mathrm{TE_{11}}$ mode. The first TM mode to propagate is then the $\mathrm{TM_{01}}$ mode. The definition of TE_{mn} and TM_{mn} mode can be found in [3]. The cutoff frequencies of $\mathrm{TE_{11}}$ and $\mathrm{TM_{01}}$ modes are listed in Table 2.3 for a circular waveguide with a radius of 15 and of 20 mm. For the third harmonic cavity, 15 mm is the iris radius of a mid-cell, while 20 mm is the iris radius of an end-cup and the radius of connecting

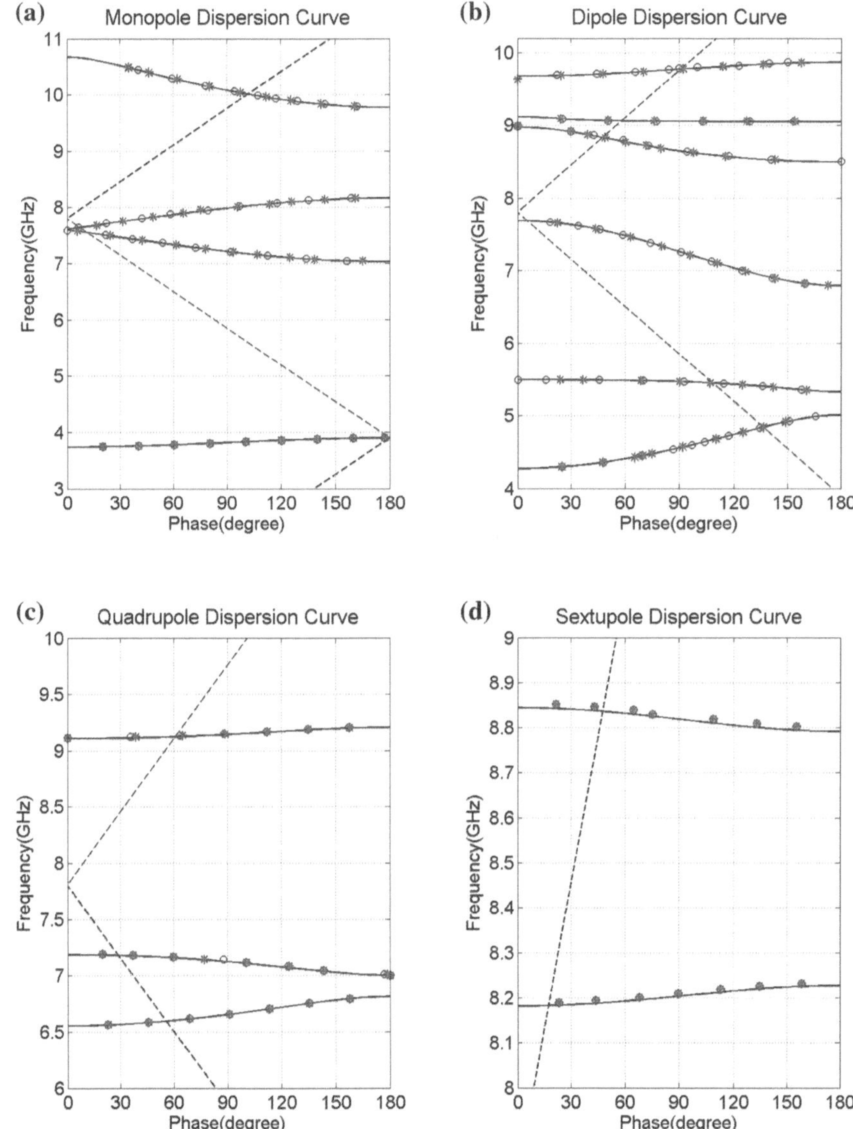

Fig. 2.9 Monopole, dipole, quadrupole and sextupole band structure (*solid curve*) of a mid-cell with periodic boundaries and the modes in an ideal 9-cell 3.9 GHz cavity. The *circles* represent modes calculated with electric (EE) boundary conditions and the *asterisks* represent magnetic (MM) boundary conditions. The light line is *dashed*

beam pipes (see Table 2.1). By choosing a beam-pipe radius larger than 1/3 of that of the 1.3 GHz cavity, the cutoff frequency is lowered so that most higher order modes propagate amongst cavities and are therefore better damped [5].

(a)

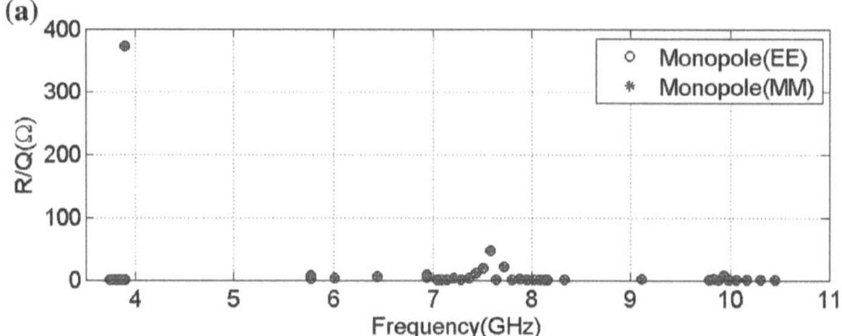

(b)

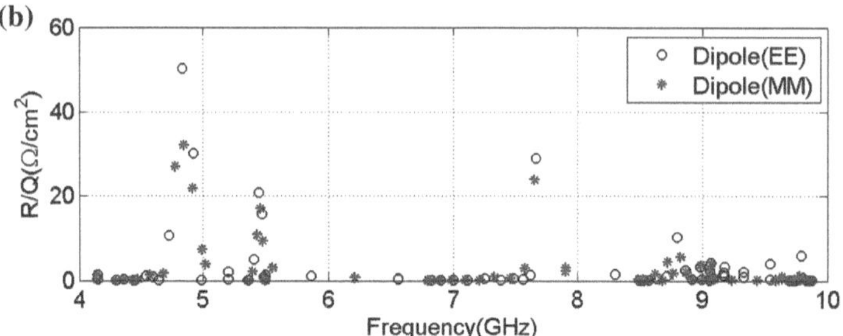

(c)

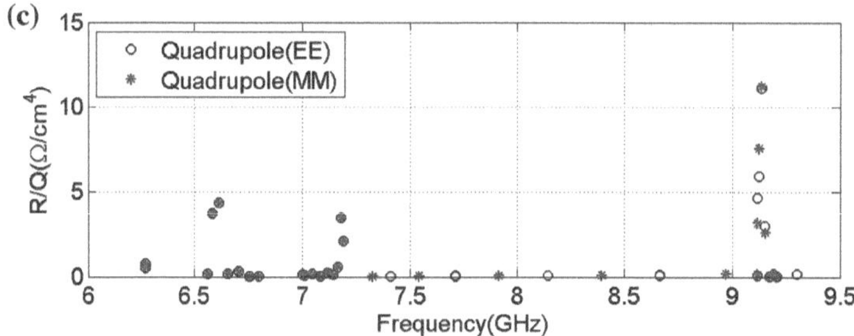

(d)

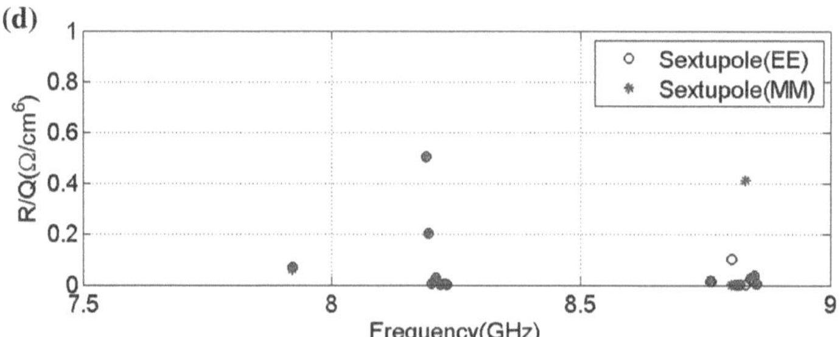

Fig. 2.10 The R/Q parameters of eigenmodes of a 9-cell third harmonic cavity versus the modal frequencies. *Circles* represent modes calculated with electric (EE) boundaries and *asterisks* with magnetic (MM) boundaries

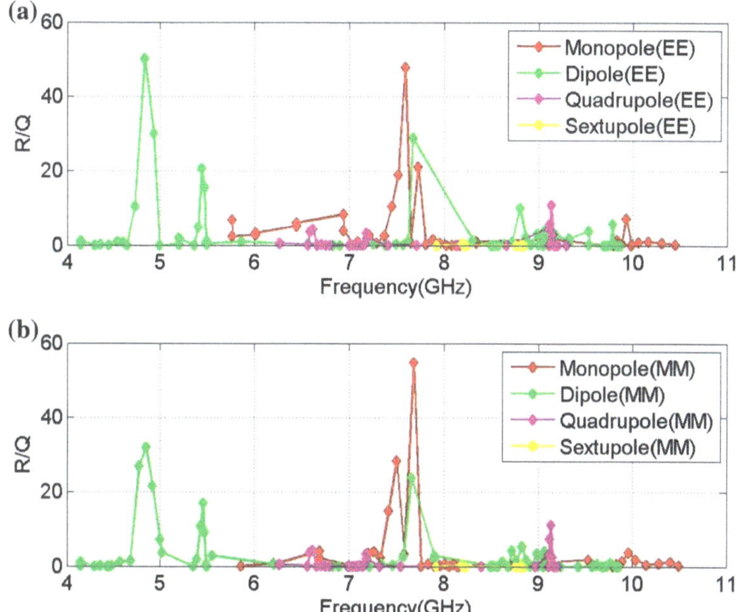

Fig. 2.11 The R/Q parameters of HOMs of a 9-cell third harmonic cavity versus the modal frequencies. The modes were calculated with electric (EE) boundaries and magnetic (MM) boundaries. The units of R/Q's are: Ω (monopole), Ω/cm^2 (dipole), Ω/cm^4 (quadrupole) and Ω/cm^6 (sextupole)

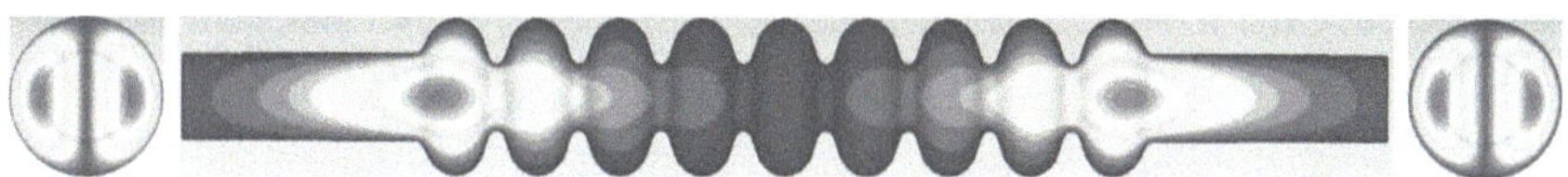

Fig. 2.12 The electric field distribution of one dipole beam-pipe mode (frequency: 4.1491 GHz, R/Q: 1.318 Ω/cm^2). Electric (EE) boundary conditions were used in the simulation. Reprinted with permission from [7]. Copyright 2012, American Institute of Physics

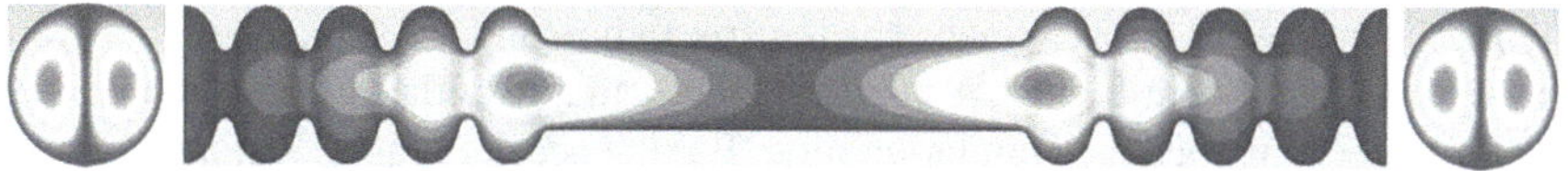

Fig. 2.13 The electric field distribution of one dipole beam-pipe mode (frequency: 4.1481 GHz, R/Q: 1.544 Ω/cm^2). Electric (EE) boundary conditions were used in the simulation

2.3 Eigenmodes of an Ideal Third Harmonic Cavity

The geometry of an ideal third harmonic cavity without couplers as modeled with CST Microwave Studio® [2] is shown in Fig. 2.5. The shape of an individual mid-cell is shown in Fig. 2.2a and the parameters are listed in Table 2.1. The end-cups have an

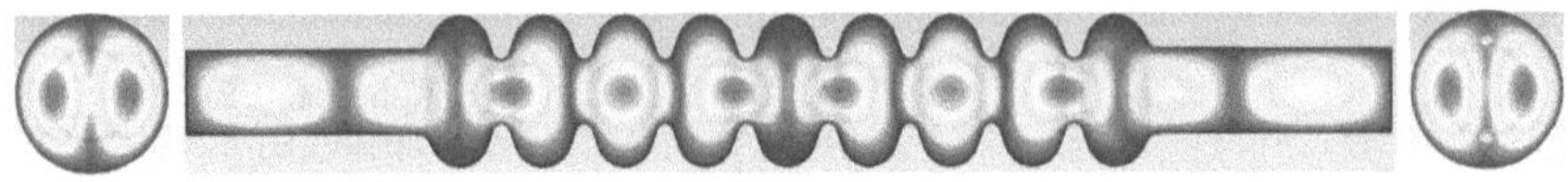

Fig. 2.14 The electric field distribution of one cavity mode from the first dipole band (frequency: 4.8327 GHz, R/Q: 50.307 Ω/cm^2). Electric (EE) boundary conditions were used in the simulation. Reprinted with permission from [7]. Copyright 2012, American Institute of Physics

Fig. 2.15 The electric field distribution of the strongest coupled cavity mode from the first dipole band (frequency: 4.8076 GHz, R/Q: 125.762 Ω/cm^2 per module). Electric (EE) boundary conditions were used in the simulation

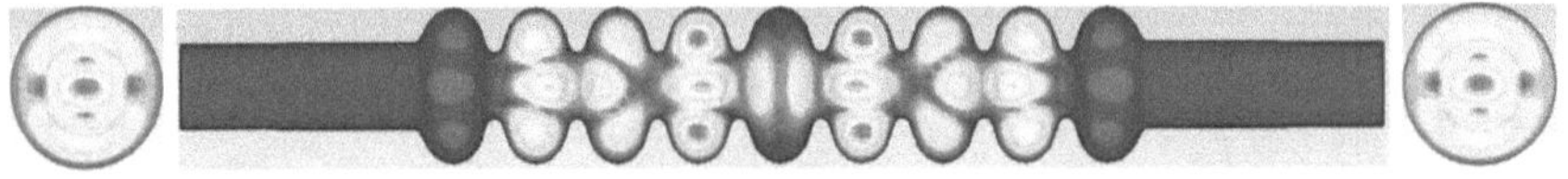

Fig. 2.16 The electric field distribution of one cavity mode from the fifth dipole band (frequency: 9.0581 GHz, R/Q: 2.171 Ω/cm^2). Electric (EE) boundary conditions were used in the simulation. Reprinted with permission from [7]. Copyright 2012, American Institute of Physics

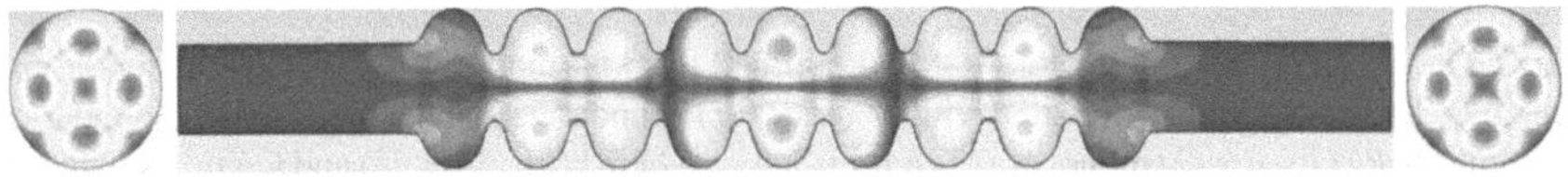

Fig. 2.17 The electric field distribution of one cavity mode from the first quadrupole band (frequency: 6.6167 GHz, R/Q: 4.358 Ω/cm^4). Electric (EE) boundary conditions were used in the simulation

increased iris radius (20 mm) and are connected with beam pipes at either ends. The simulations were conducted with Eigenmode Solver of CST Microwave Studio®. A solver accuracy of 10^{-6} in terms of the eigensystem's relative residual was used. The cavity geometry was approximated by hexahedral mesh cells. As shown in Fig. 2.6, the mesh lines were chosen such that the iris radius and the equator radius were exactly matched by mesh lines. A quarter of the structure with symmetry planes was used in order to reduce the simulation time. For the accelerating mode, a maximum mesh step of 1.1 mm, corresponding to approximately 2.1 million mesh cells for a quarter of the structure, was used. Electric (EE) boundary conditions were used both on the surface and the beam pipe ends.

The electric field of the accelerating mode (3.9 GHz, π mode [5]) is shown in Fig. 2.7a, b. The unit of R/Q is [Ω per cavity]. Throughout this report, R/Q is calculated "per cavity" and is often omitted from the unit unless explicitly otherwise stated. The longitudinal component of the electric field of the accelerating mode on the cavity axis is shown in Fig. 2.7c. The field flatness can be calculated as [6]

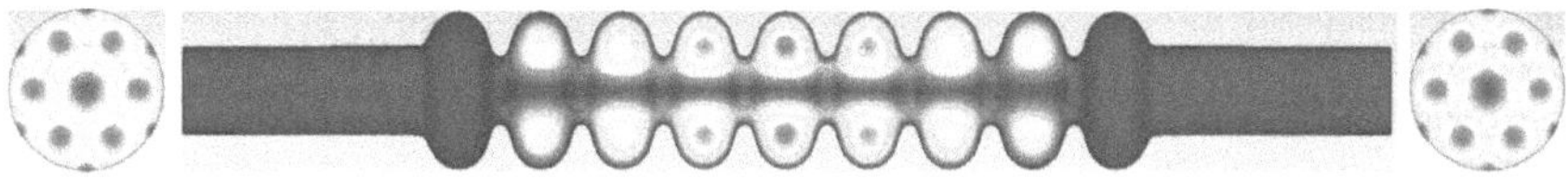

Fig. 2.18 The electric field distribution of one cavity mode from the first sextupole band (frequency: 8.1894 GHz, R/Q: 0.506 Ω/cm^6). Electric (EE) boundary conditions were used in the simulation

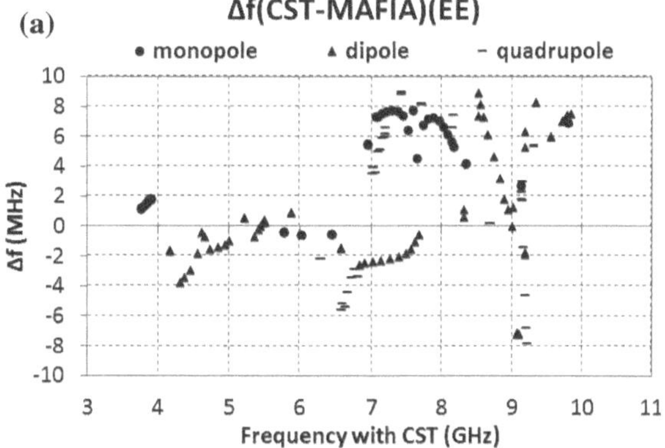

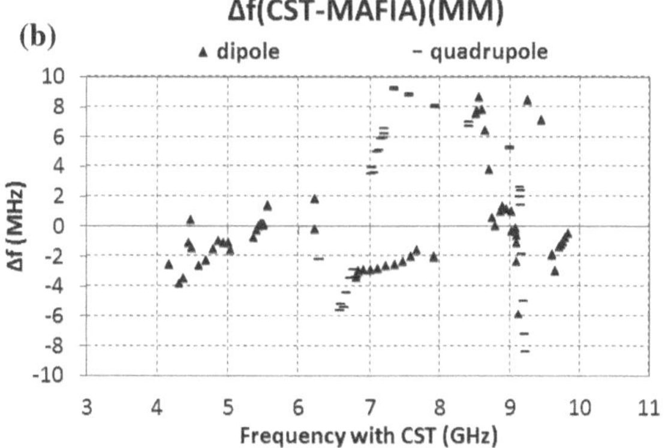

Fig. 2.19 Frequency differences of modes simulated with CST® and MAFIA®. Δf is calculated as $\Delta f = f_{CST} - f_{MAFIA}$

$$\text{field flatness} = \frac{(E_{cmax} - E_{cmin})}{\frac{1}{N}\sum_{i=1}^{N} E_{ci}} \times 100\,\%, \tag{2.5}$$

where E_{cmax} and E_{cmin} are the maximum and minimum peak axial electric fields in the multi-cell cavity, E_{ci} is the peak axial electric field in the ith cell, N is the total number of cells in the cavity. In our case, $N = 9$. The field flatness is calculated to be 3.5 % for the accelerating mode.

The phase advance per cell can be calculated using the electric field determined from the simulations. Based on Eq. 2.1 for periodic structures, the phase advance per cell can be derived according to the Floquet periodicity conditions [4]:

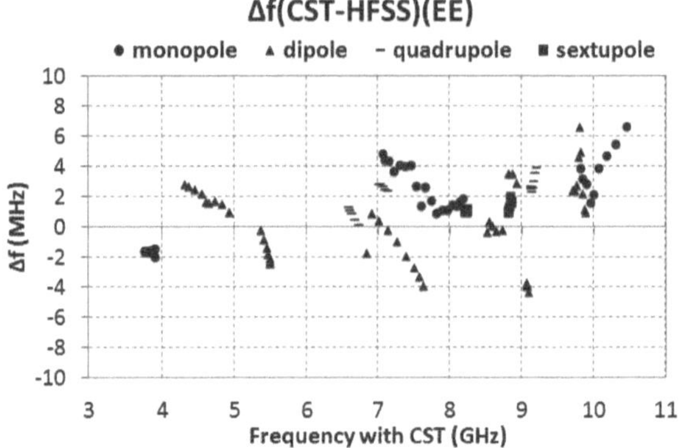

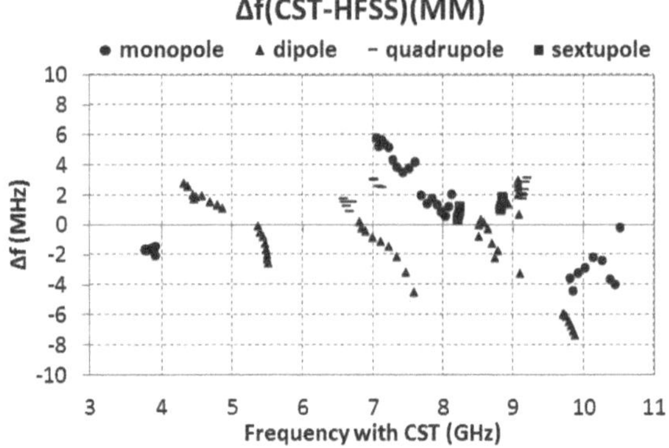

Fig. 2.20 Frequency differences of modes simulated with CST® and HFSS®. Δf is calculated as $\Delta f = f_{CST} - f_{HFSS}$

$$E_M = E_z(r, z), \tag{2.6}$$

$$E_L = E_M e^{-i\phi} = E_z(r, z - L), \tag{2.7}$$

$$E_R = E_M e^{i\phi} = E_z(r, z + L), \tag{2.8}$$

$$E_L + E_R = 2E_M cos(\phi), \tag{2.9}$$

where $E_z(r, z)$ is the longitudinal electric field obtained from the simulations, E_L, E_M and E_R are defined as illustrated in Fig. 2.8. The phase advance per cell can be calculated as

$$\phi = arccos\left(\frac{E_L + E_R}{2E_M}\right). \tag{2.10}$$

The phase advance per cell calculated in the middle of the cavity for the accelerating mode is 177 degrees. This reflects the field flattness that is adjusted by the geometry of the end-cells.

Equation 2.10 has been used to calculate the phase advance per cell for all modes which have been simulated. The results for monopole, dipole, quadrupole and sextupole bands are shown in Fig. 2.9 along with dispersion curves of the mid-cell presented in Sect. 2.1. Beside the electric (EE) boundaries, another set of simulations with magnetic (MM) boundaries set on both beam pipe ends has also been presented. At both ends of the cavity, the longitudinal electric field is zero for the MM boundaries, while the transverse electric field is zero for the EE boundaries. These are used for the study of field distributions of the eigenmodes. The frequencies of the first monopole band are below the cutoff frequency of the beam pipe (see Table 2.3), therefore modes in this band do not depend on the boundary conditions. This can be seen in Fig. 2.9a as the overlap of asterisks and circles. Some dipole modes in the fifth dipole band and the first two modes in the first dipole band are trapped within the cavity, which explains the consistency of results from EE and MM boundary conditions in Fig. 2.9b. Large deviations for different boundary conditions can be clearly seen in other dipole bands as they are propagating amongst cavities. Figure 2.10a–d show the R/Q value versus the frequency of each mode for both EE and MM boundary conditions. The coupling strengths for all HOMs beyond the fundamental band are shown in Fig. 2.11a and b.

In addition to cavity modes shown in the bands, there are also beam-pipe modes, whose electromagnetic energy mainly deposits in beam pipes and end-cells of the cavity. These modes are trapped within both beam-pipe ends of the structure. One of these modes is shown in Fig. 2.12. The same mode has also been simulated with a structure shown in Fig. 2.13. The dipole character of this mode can be seen clearly in the projection on the transverse plane in the middle of each end-cell.

The propagating feature of one dipole cavity mode can be seen in Figs. 2.14 and 2.15. The mode can couple to adjacent cavities through attached beam pipes, and also has strong coupling to the beam represented by the large R/Q value.

One trapped cavity mode from the fifth dipole band is shown in Fig. 2.16. Compared with other trapped modes in this band, this mode has stronger coupling to the beam (larger R/Q value).

One quadrupole mode and one sextupole mode are also shown in Figs. 2.17 and 2.18. The R/Q values are in general small for these modes.

Compared to the eigenmode simulations obtained by using MAFIA® [8] and HFSS® [9], the frequencies of modes are shifted. A direct comparison between CST® and MAFIA® is shown in Fig. 2.19, while a comparison between CST® and HFSS® is shown in Fig. 2.20. The MAFIA® simulation results are from Ref. [1] while the HFSS® simulations are from Ref. [10]. The differences are within 10 MHz for both boundary conditions from both simulation codes. This is comparable to the frequency shift of HOMs in the real cavity from their eigenmode values due to asymmetric cavity structure caused by the couplers and fabrication tolerances. These are described in Chap. 3.

A list of simulated eigenmodes for an ideal third harmonic cavity is shown in Appendix B. The electric field distributions for the first monopole band, the first two dipole bands and the fifth dipole band are also shown in Appendix B for both electric (EE) and magnetic (MM) boundary conditions. Extensive electric field distributions for monopole, dipole, quadrupole and sextupole eigenmodes up to 10 GHz are described in Ref. [11] with both electric (EE) and magnetic (MM) boundary conditions. Simulations of a four-cavity third harmonic module with couplers and beam-pipe bellows can be found in Ref. [12–16].

Having the knowledge of the band structure of the third harmonic 3.9 GHz cavity and field distributions of various modes from simulations, the experimental characterization of HOMs in the cavity/module both with and without beam-excitations are discussed in the Chap. 3.

References

1. T. Khabibouline et al., Higher order modes of a 3rd harmonic cavity with an increased end-cup Iris. TESLA-FEL Report: TESLA-FEL 2003-01 (2003)
2. C.S.T. Microwave, Studio®. Ver., CST AG, Darmstadt, Germany (2011)
3. D.M. Pozar, *Microwave Engineering*, third edn. Wiley, New York (2005)
4. R. Wanzenberg, Monopole, dipole and quadrupole passbands of the TESLA 9-cell cavity. TESLA Report: TESLA 2001-33 (2001)
5. J. Sekutowicz, R. Wanzenberg, W.F.O. Mueller, T. Weiland, A design of a 3rd harmonic cavity for the TTF 2 photoinjector. TESLA-FEL Report: TESLA-FEL 2002-05 (2002)
6. S. An, H. Wang, Tuner effect on the field flatness of SNS superconducting RF cavity. JLAB Note: JLAB-TN-03-043 (2003)
7. P. Zhang, N. Baboi, R.M. Jones, I.R.R. Shinton, T. Flisgen, H.W. Glock, A study of beam position diagnostics using beam-excited dipole modes in third harmonic superconducting accelerating cavities at a free-electron laser. Rev. Sci. Instrum. **83**, 085117 (2012)
8. MAFIA Release 4. CST AG, Darmstadt, Germany
9. ANSYS®HFSS. Release 11.2, ANSYS Inc., USA
10. I.R.R. Shinton et al., Compendium of eigenmodes in third harmonic cavities for FLASH and the XFEL. DESY Report: DESY 12-053 (2012)
11. P. Zhang, N. Baboi, R.M. Jones, Eigenmode simulations of third harmonic superconducting accelerating cavities for FLASH and the European XFEL. DESY Report: DESY 12-101 (2012)
12. I.R.R. Shinton et al., Higher order modes in third harmonic cavities for XFEL/FLASH, in *Proceedings of IPAC'10, (Kyoto, Japan)*, pp. 3007–3009 (2010)
13. I.R.R. Shinton et al., Higher order modes in third harmonic cavities at FLASH, in *Proceedings of Linear Accelerator Conference LINAC2010, (Tsukuba, Japan)*, pp. 785–787 (2010)
14. I.R.R. Shinton et al., Higher order modes in coupled cavities of the FLASH module ACC39, in *Proceedings of IPAC2011, (San Sebastian, Spain)*, pp. 2301–2303 (2011)
15. T. Flisgen et al., A concatenation scheme for the computation of beam excited higher order mode port signals, in *Proceedings of IPAC2011, (San Sebastian, Spain)*, pp. 2238–2240 (2011)
16. I.R.R. Shinton, R.M. Jones, Z. Li, P. Zhang, Simulations of higher order modes in the ACC39 module of FLASH, in *Proceedings of IPAC2012, (New Orleans, Louisiana, USA)*, pp. 1900–1902 (2012)

Chapter 3
Measurements of HOM Spectra

In order to understand the HOM spectra of third harmonic cavities, modal characterization is needed at all stages after the fabrication of cavities: transmission spectra between two HOM couplers measured for each isolated single cavity are presented in Sect. 3.1, module-based transmission measurements in Sect. 3.2 and beam-excited HOM measurements in Sect. 3.3. The signals were measured from all eight HOM couplers. Not all measured signals are shown in this chapter. The selection is based on a representative sample of the typical behavior of the modes. HOM signals measured from all eight HOM couplers can be found in Ref. [1]. In particular, trapped modes and inter-cavity coupled modes are focused on with a view to understanding their application to HOM-based beam diagnostics.

3.1 Transmission Spectra of an Isolated Single Cavity

Third harmonic cavities were mounted on the test stand and cooled down at Fermilab after fabrication [2]. The RF transmission spectrum (S_{21}) was subsequently measured for each isolated single cavity[1] as shown schematically in Fig. 3.1. A typical spectrum is shown in Fig. 3.2a along with simulations of an ideal single cavity without couplers (see Chap. 2). The band structure of a single cavity is depicted in terms of monopole, dipole, quadrupole and sextupole modes from 3.7 to 10 GHz frequency span. As it will be explained later, regions of particular interests are respectively the fundamental band (Fig. 3.2b), the first two dipole bands (Fig. 3.2d) and the fifth dipole band (Fig. 3.2c). The nine modes in the fundamental band can be identified in Fig. 3.2b. The accelerating mode is the last peak at 3.9 GHz (π mode as shown in Chap. 2). Unlike the simulated cavity, the actual cavity has couplers, which breaks the symmetry of the structure. This accounts for the differences between simulations and measurements. A direct simulation of a real cavity with couplers is computationally very expensive in terms of time and resources, thus alternative techniques were

[1] Data are kindly provided by T. Khabibouline from Fermilab.

P. Zhang, *Beam Diagnostics in Superconducting Accelerating Cavities*, Springer Theses, DOI: 10.1007/978-3-319-00759-5_3,

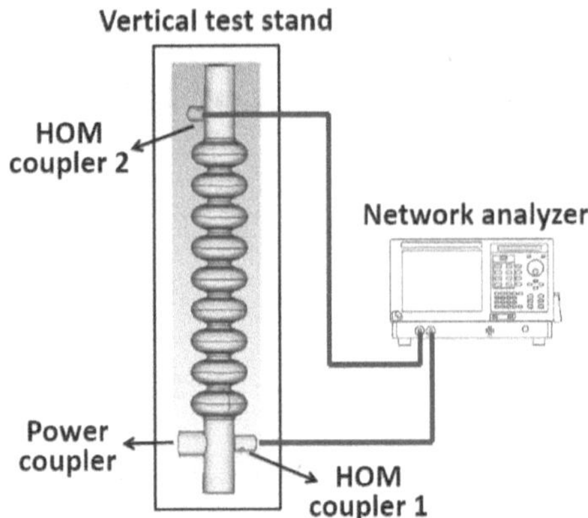

Fig. 3.1 The schematic setup of the single cavity RF measurement

applied [3–6]. In addition, other sources such as fabrication errors and cavity tuning can also contribute to the differences. The dipole modes are not readily identifiable as shown in Figs. 3.2c, d.

For comparison, a typical beam-excited spectrum of the first dipole band of the TESLA 1.3 GHz cavity [7] is shown in Fig. 3.3. Here one can identify nine peaks, which are clearly separated from each other. According to simulations [8], peaks marked as #6 and #7 in Fig. 3.3 are identified as modes which have strong coupling to the beam. Therefore, these modes are more sensitive to the transverse beam offset, which will lead to a better resolution when used for position diagnostics. Mode #6 at approximately 1.7 GHz was used for HOMBPM in 1.3 GHz cavities [9]. The R/Q of this mode is 5.54 Ω/cm^2 [8]. The modal spectrum is more complicated for the 3.9 GHz cavity (see Fig. 3.2d). Mode identification is by no means straightforward.

In general, each peak in Fig. 3.2 can be fit to a Lorentzian distribution [11]:

$$y = \frac{y_0}{1 + \left(\frac{f - f_0}{\Delta f}\right)^2}. \tag{3.1}$$

Here y_0 is the mode amplitude, f_0 is the center frequency and Δf is the half-width at half-amplitude (HWHM). In order to identify peaks in a complex spectrum, all modes in a dipole band are fit simultaneously rather than fitting individual peaks. The frequency f_0 and the quality factor Q ($Q = f_0/(2\Delta f)$) are then obtained for each mode. Figure 3.4 shows the fitting results of the first two dipole bands using the commercial code PeakFit [12]. On the top half of each figure, the original spectrum is plotted in dotted line while the fitted one in solid line. On the bottom half of each figure, each peak denotes a mode with a fitted frequency marked on top. Hidden peaks are revealed. The goodness of fit is measured by the coefficient of determination r^2 ($r^2 = 1$, perfect fit; $r^2 = 0$, poor fit; see Appendix A.3). The Q of each mode is shown in Fig. 3.5 along with the coupling strength to the beam for each mode from simulations (described in Chap. 2.3). The Q_{ext} is also shown in Fig. 3.5 for each mode

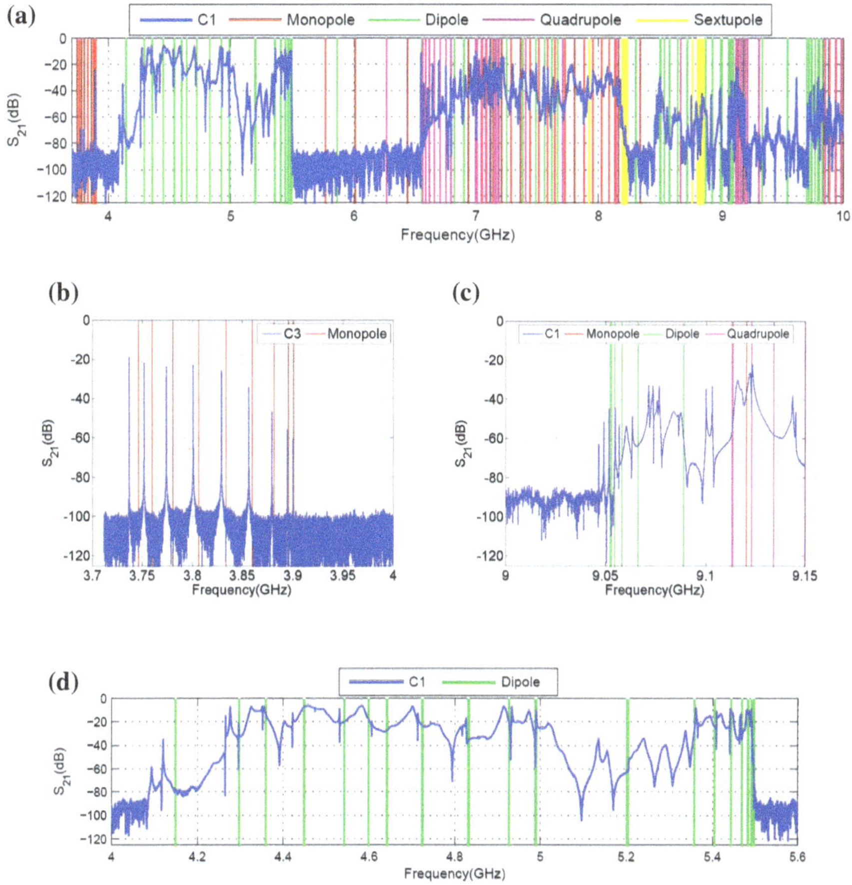

Fig. 3.2 **a** Measured across C1 from C1H1 to C1H2. **b** The fundamental band. **c** The fifth dipole band. **d** The first two dipole bands.Typical transmission spectrum (S_{21}) of a single isolated cavity. The vertical lines indicate simulation results of eigenmodes (Appendix B.1)

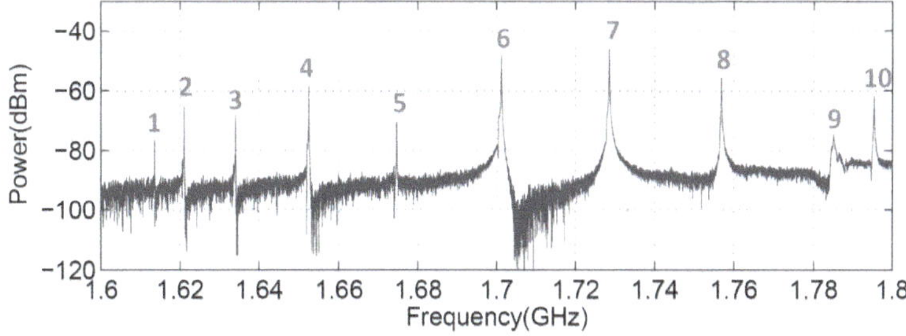

Fig. 3.3 The beam-excited spectrum of the first dipole band of the TESLA 1.3 GHz cavity. The spectrum was measured from HOM coupler C6H2 of the ACC1 module (see Fig. 1.3) [10]

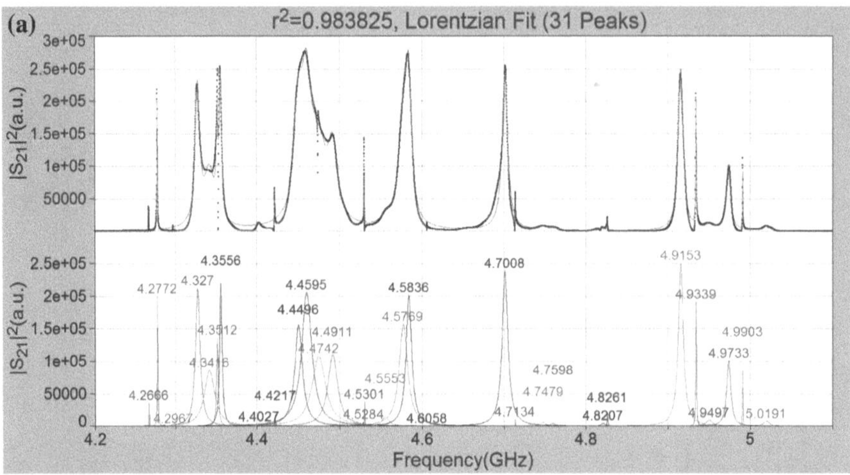

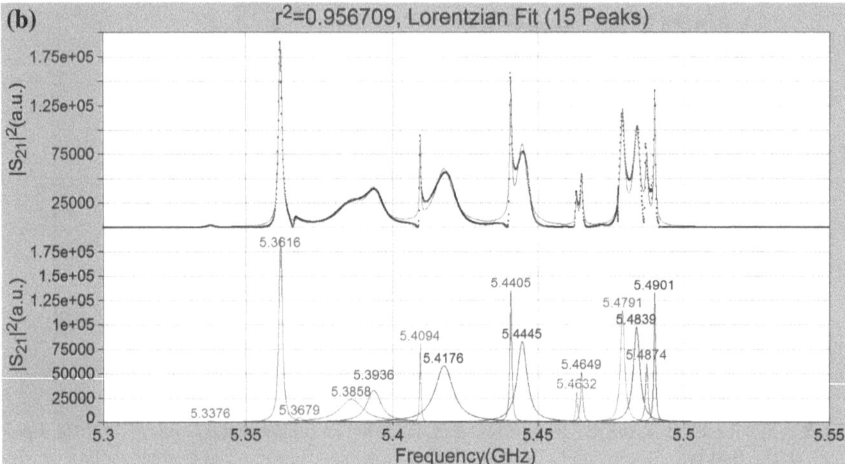

Fig. 3.4 **a** The first dipole band. **b** The second dipole band. Fit of the first two dipole bands of C1 (from the isolated single cavity measurement) as Lorentzian distributions

simulated on a cavity with power and HOM couplers [13]. The total damping of a cavity mode is not only determined by finite resistance of the cavity wall (denoted by Q) but also by the external quality factor, Q_{ext}, which characterizes the coupling to HOM couplers. The polarization split of dipole modes and the frequency shift from ideal simulations can be observed. By comparing the modal frequencies between simulations and measurements, modes can be approximately identified, but this is by no means as straightforward or precise as it was with the TESLA-style cavities. The fitting results of the fifth dipole band are shown in Figs. 3.6 and 3.7. In reality the cavities are connected through beam pipes within a module. The connecting beam pipes are above cut-off of the TE/TM dipole modes. It is important to characterize these coupling modes within a module. This is described in the next section.

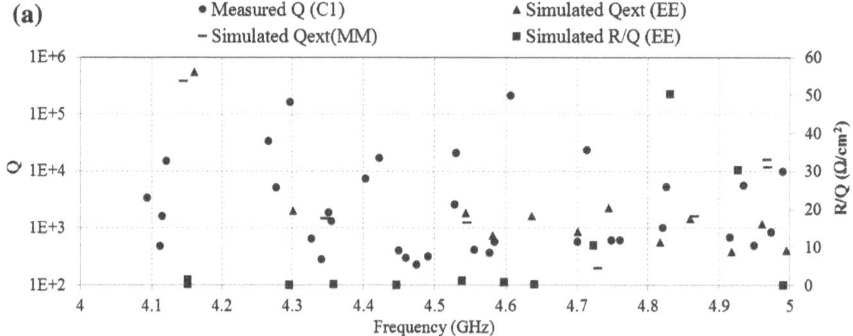

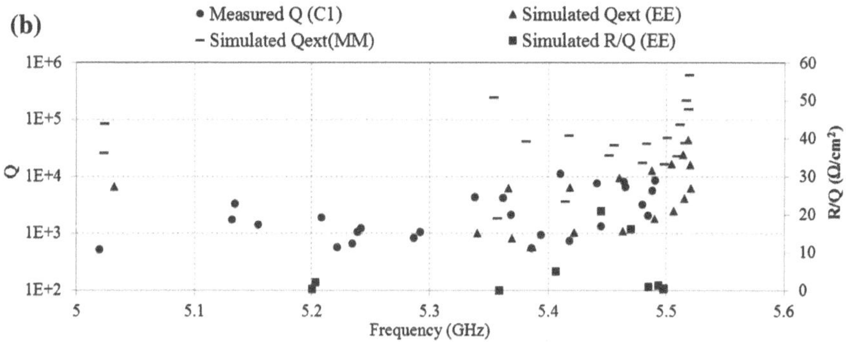

Fig. 3.5 **a** The first dipole band. **b** The second dipole band. **c** Simulations and single cavity measurements of the dipole beam-pipe modes and the first two dipole bands of C1 corresponding to Fig. 3.4

3.2 Module-Based Transmission Spectra

After the assembly of all four cavities into the cryo-module ACC39, the module was installed in the Cryo-Module Test Bench (CMTB) at DESY. RF transmission measurements were conducted in the CMTB tunnel[2] as shown schematically in Fig. 3.8. The measurements were repeated after the installation in FLASH from the HOM board rack outside the tunnel.[3] HOM signals are stronger in CMTB than in FLASH because additional amplifiers were used and the cables are shorter. The two measurements are otherwise similar. Therefore CMTB results are shown regarding most of the module-based measurements without beam-excitation. The results of measurements at FLASH can be found in Ref. [1]. From the recorded spectra, monopole, dipole and quadrupole bands are identified. Instead of usual discrete modes in

[2] Data are kindly provided by T. Khabibouline from Fermilab.

[3] H.W. Glock from University of Rostock led the measurements.

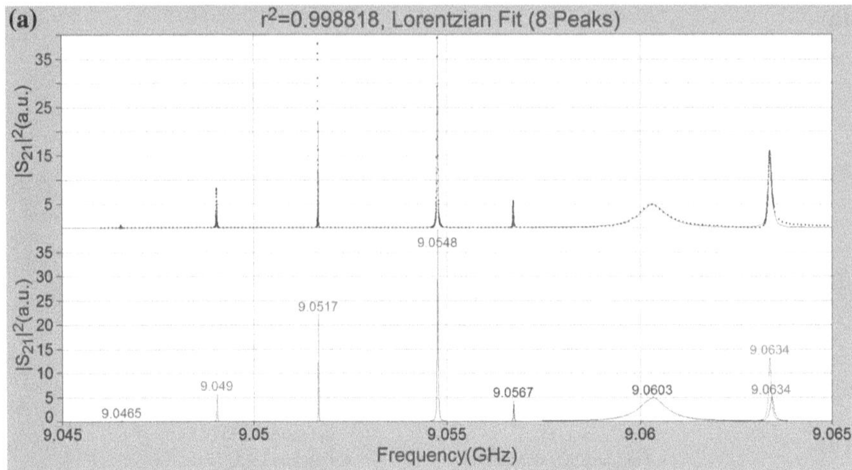

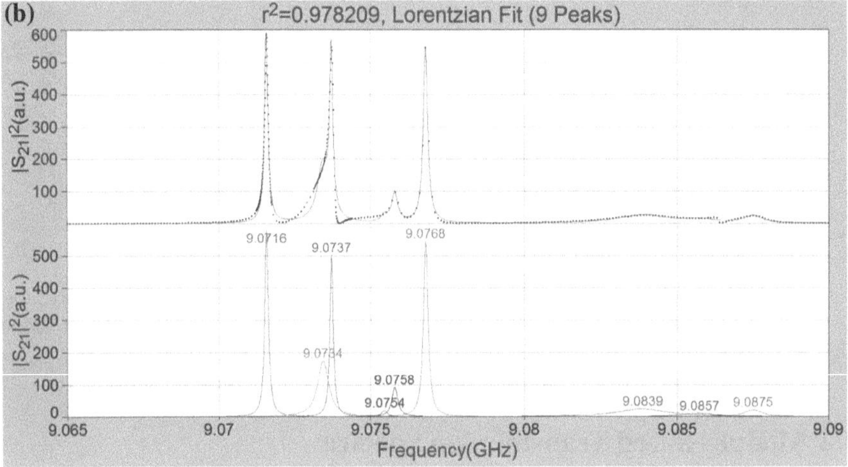

Fig. 3.6 **a** part 1. **b** part 2. Fit of the fifth dipole band of C1 (from isolated single cavity measurements) as Lorentzian distributions

the isolated single cavity spectrum described in Chap. 3.1, more modes arise due to the coupling of inter-connected cavities. This coupling can be clearly seen in Fig. 3.9 where the S_{21} parameter of an isolated cavity is compared with the one measured for the same cavity at CMTB.

In order to characterize the inter-cavity couplings of these modes, transmission spectrum was measured along the entire four-cavity string (from C1H2 to C4H2), and then compared with the spectra measured between upstream and downstream couplers of each cavity within the cryo-module. As expected from simulations (Chap. 2.3), most HOMs couple to adjacent cavities through attached beam pipes. This can be observed in Fig. 3.10, where only a small portion of modes is below the

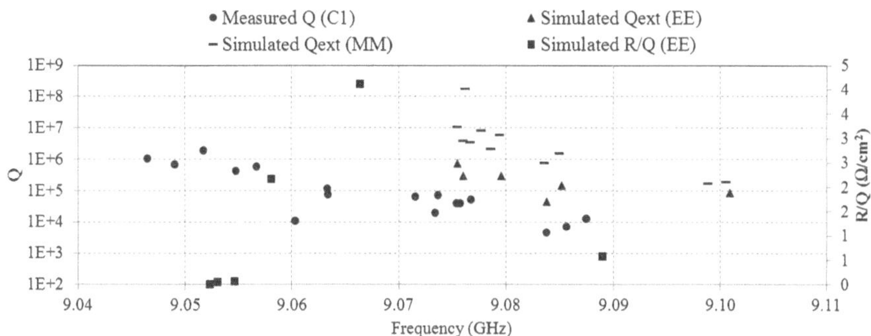

Fig. 3.7 Simulations and single cavity measurements of the fifth dipole band of C1 corresponding to Fig. 3.6

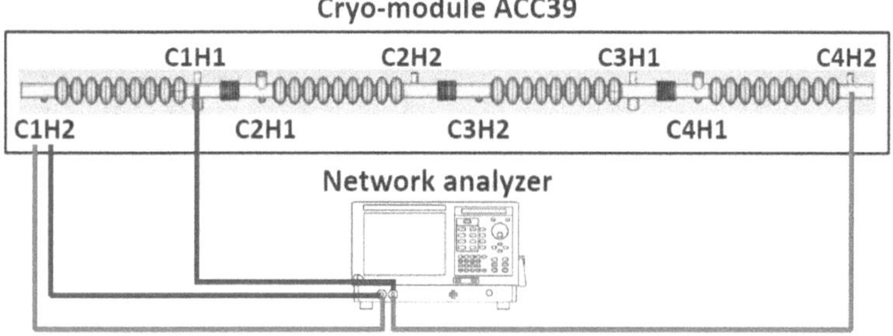

Fig. 3.8 The schematic setup of the module-based RF measurement

cutoff frequencies of attached beam pipes (peaks present only in the cavity C1 spectrum but absent from the string spectrum). As described in Chap. 2.2, the theoretical cutoff frequency of the beam pipe is 4.39 GHz for dipole TE_{11} modes, while the measurement shows an even lower value at approximately 4.25 GHz. This might be due to the asymmetric geometry due to couplers and to fabrication tolerances. Dipole beam-pipe modes at approximately 4.1 GHz and cavity modes in the fifth dipole band at approximately 9.05 GHz are clearly visible as localized within each beam pipe or cavity. Simulations presented in Chap. 2.3 also show the non-propagating characteristics of these modes.

3.3 Beam-Excited HOM Spectra

The beam-excited spectra measurement at FLASH was conducted using a Tektronix Oscilloscope (scope) with a bandwidth of 6 GHz and a Tektronix Real-time Spectrum Analyzer (RSA) [15, 16]. As shown in Fig. 3.11, HOM signals were taken from both HOM couplers of all four cavities at ACC39 HOM patch panel outside the tunnel. A

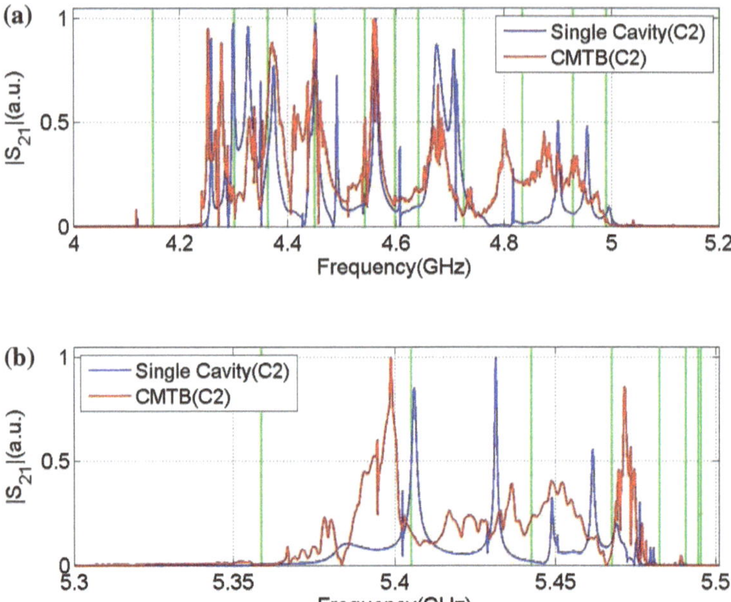

Fig. 3.9 **a** The first dipole band of C2. **b** The second dipole band of C2. Spectra of C2 from the isolated single cavity measurement (*blue*) and CMTB (*red*). The vertical lines are from simulations

10 dB external attenuator was connected to each HOM coupler to reduce the power of the beam-excited signals radiated to the coupler and thus protecting the devices. A RF multiplexer was used to connect one coupler to either the scope or the RSA during each measurement. Both the scope and the RSA were triggered by a 10 Hz RF signal delivered by the FLASH timing system, synchronized with the beam pulse. We were running FLASH with one bunch per beam pulse in a repetition rate of 10 Hz. The bunch charge was approximately 0.5 nC. A MATLAB [17] script was used to control the RF multiplexer, the scope and the RSA as well as the data recording to the local computer hard disk. Time-domain waveforms and real-time spectra were recorded. Each waveform was sampled with 20 GS/s, and 200,000 points were recorded in a time window of 10 μs. Example waveforms from the upstream and the downstream couplers of C2 are shown in Fig. 3.12.

The RSA has the capability to capture the HOM signal excited by a single bunch in a large bandwidth with a high frequency resolution. In our case, the acquisition bandwidth was set to 60 MHz with an acquisition length of 100 μs. A frequency step of 10 kHz and a resolution bandwidth of 22.5 kHz were used. The frequency span was set to 50 MHz. Starting from 4.025 GHz, the center frequency was raised in a step of 50 MHz. Thus the spectral ranges were consecutively recorded. Each 50 MHz slice of the spectrum corresponds to a single bunch excitation. Figure 3.13 shows the HOM spectrum ranging from 4 to 9.5 GHz measured from coupler C1H1. Figure 3.14

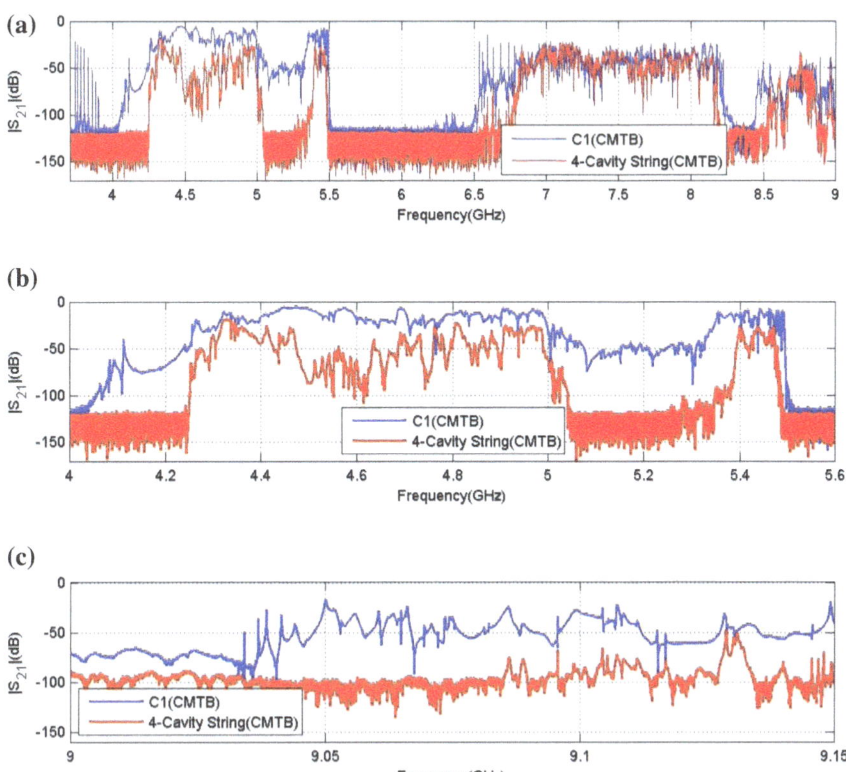

Fig. 3.10 **a** Spectrum from 3.7 to 9 GHz. **b** The first two dipole bands. **c** The fifth dipole band. Couplings of inter-connected cavities. The measurements were made at CMTB, while the spectra measured across C1 (from C1H1 to C1H2) is in *blue*, and the spectra measured across the entire four-cavity string (from C1H2 to C4H2) is in *red*. Reprinted with permission from [14]. Copyright 2012, American Institute of Physics

shows the beam-excited spectrum of the fundamental band of one cavity compared with transmission spectrum measured at CMTB without beam-excitation. After a Fast Fourier Transform (FFT) applied on the time-domain waveform (Fig. 3.12), HOM signals measured by scope and RSA are compared in Fig. 3.15. They show very good consistency. The spectrum of the fifth dipole band is presented in Fig. 3.16 (without the 10 dB external attenuator).

After this series of HOM spectra measurements, the band structure of modes has been characterized. Two categories of dipole modes of the third harmonic cavities have been distinguished. Modes in one category are localized in each cavity or segment of beam pipe, while in another category are propagating through the entire four-cavity module. Localized dipole modes consist of beam-pipe modes at approximately 4 GHz and some cavities modes in the fifth dipole band at approximately 9 GHz. These modes only possess small R/Q values according to simulations

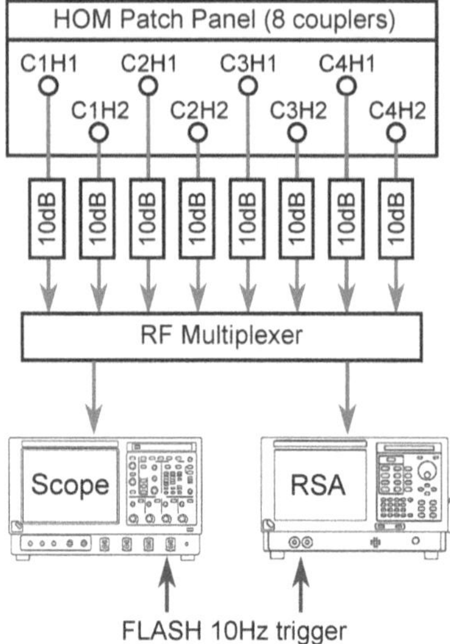

Fig. 3.11 Schematic setup for the beam-excited HOM measurements. The 10 dB external attenuators have been removed during the measurement of the fifth dipole band. Reprinted with permission from [14]. Copyright 2012, American Institute of Physics

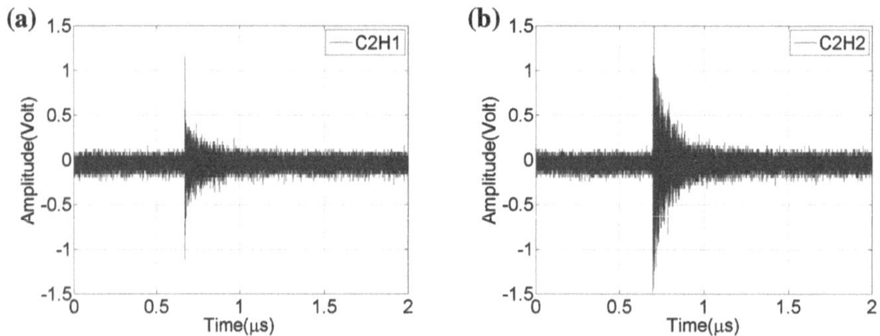

Fig. 3.12 **a** C2H1. **b** C2H2. Waveforms excited by a single electron bunch measured from the upstream and the downstream HOM coupler of C2

(Chap. 2.3). Coupled dipole modes with large R/Q values consist of cavity modes in the first two dipole bands within 4.5–5.5 GHz. These two categories of modes are of particular interests for beam position diagnostics. Their dependencies on transverse beam offsets will be studied in Chap. 5. Data analysis techniques used to extract beam position information from HOM signals must be employed and these techniques are explained in the Chap. 4.

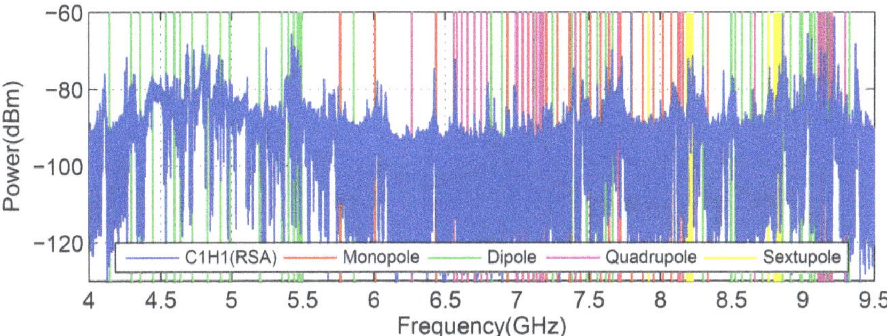

Fig. 3.13 Beam-excited spectrum measured from HOM coupler C1H1. The vertical lines indicate simulation results of an ideal cavity

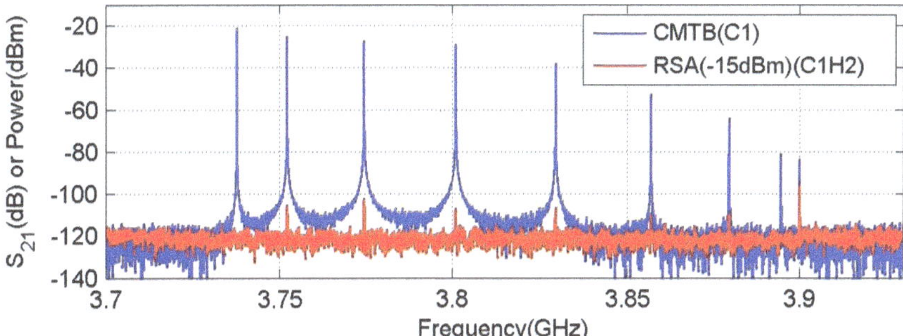

Fig. 3.14 Beam-excited spectrum of the fundamental band (15 dB subtracted, *red*) measured from HOM coupler C1H2 compared with transmission spectrum measured at CMTB without beam-excitation (*blue*)

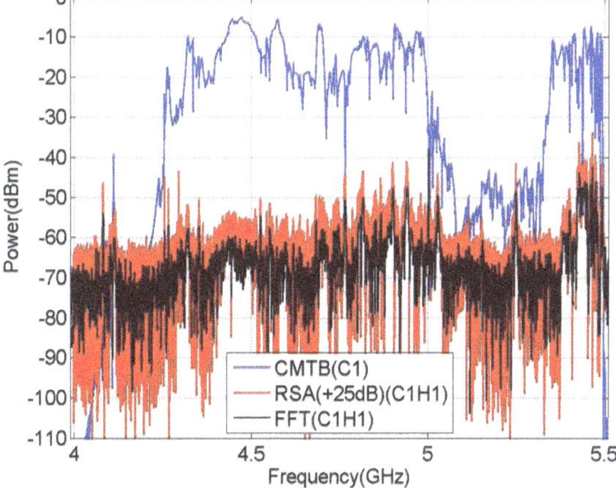

Fig. 3.15 Beam-excited spectrum of the first and second dipole band measured from HOM coupler C1H1 using RSA (25 dB added, *red*) and scope (after FFT, *black*). The transmission spectrum measured at CMTB without beam-excitation is in *blue*

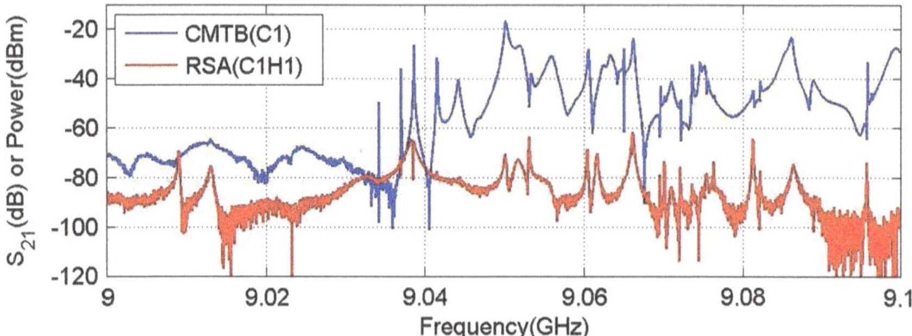

Fig. 3.16 Beam-excited spectrum of the fifth dipole band (*red*) measured from HOM coupler C1H1 compared with transmission spectrum measured at CMTB without beam-excitation (*blue*). The 10 dB attenuators have been removed during this measurement

References

1. P. Zhang, N. Baboi, R.M. Jones, Higher order mode spectra and the dependence of localized dipole modes on the transverse beam position in third harmonic superconducting cavities at FLASH. DESY Report: DESY 12–109 (2012).
2. T. Khabibouline. (private communication).
3. I.R.R. Shinton et al., *Higher order modes in third harmonic cavities at FLASH, in Proceedings of Linear Accelerator Conference LINAC2010* (Tsukuba, Japan, 2010), pp. 785–787
4. I.R.R. Shinton et al., Higher order modes in coupled cavities of the FLASH module ACC39. in Proceedings of IPAC2011, (San Sebastian, Spain, 2011), pp. 2301–2303.
5. T. Flisgen et al., *A concatenation scheme for the computation of beam excited higher order mode port signals, in Proceedings of IPAC2011* (San Sebastian, Spain, 2011), pp. 2238–2240
6. I.R.R. Shinton, R.M. Jones, Z. Li, P. Zhang, *Simulations of higher order modes in the ACC39 module of FLASH, in Proceedings of IPAC2012* (New Orleans, Louisiana, 2012), pp. 1900–1902
7. J. Sekutowicz, *Multi-cell Superconducting Structures for High Energy e+e- Colliders and Free Electron Laser Linacs*, 1st edn. (Warsaw University of Technology Publishing House, Warsaw, 2008)
8. R. Wanzenberg, Monopole, dipole and quadrupole passbands of the TESLA 9-cell Cavity. TESLA Report: TESLA 2001–33 (2001).
9. S. Molloy et al., High precision superconducting cavity diagnostics with higher order mode measurements. Phys. Rev. ST Accel. Beams **9**, 112801 (2006)
10. TESLA Technical Design Report. DESY Report: DESY 01–011 (2001).
11. S.S.M. Wong, Computational Methods in Physics and Engineering, ch. 6–1, 2nd edn. (World Scientific Publishing Co., Singapore, 1997), p. 245.
12. PeakFit. Ver. 4.12, Systat Software Inc., San Jose, California, USA.
13. T. Khabibouline et al., Higher order modes of a 3rd harmonic cavity with an increased end-cup iris. TESLA-FEL Report: TESLA-FEL 2003–01 (2003).
14. P. Zhang, N. Baboi, R.M. Jones, I.R.R. Shinton, T. Flisgen, H.W. Glock, A study of beam position diagnostics using beam-excited dipole modes in third harmonic superconducting accelerating cavities at a free-electron laser. Rev. Sci. Instrum. **83**, 085117 (2012)
15. Fundamentals of Real-Time Spectrum Analysis, Tektronix Application Note (2010).
16. P. Zhang et al., *First beam spectra of SC third harmonic cavity at FLASH, in Proceedings of Linear Accelerator Conference LINAC2010* (Tsukuba, Japan, 2010), pp. 782–784
17. MATLAB. Ver. R2011b, The MathWorks Inc., USA. http://www.mathworks.com/

Chapter 4
Analysis Methods for Beam Position Extraction from HOM

Beam position information is concealed in HOMs. Signals are well defined in conventional BPMs, thus the beam position can be determined in a straightforward way [1]. However, HOMs in third harmonic 3.9 GHz cavities are much more complex as discussed in Chap. 3, therefore advanced data analysis methods are required in order to extract the beam position information from HOMs effectively [2]. These methods are explained in detail in this chapter. Data samples used to describe the methods are focused on trapped cavity modes from the fifth dipole band ranging from 9.05 to 9.08 GHz. The spectra were measured from HOM coupler C1H1 using the RSA for various beam trajectories. The detailed measurement scheme will be described in Chap. 5.

4.1 Data Preparation

Prior to extracting beam position information from HOM signals, we first define our problem in this section. All 184 HOM spectra obtained are shown in Fig. 4.1a. The corresponding transverse beam positions at the center of cavity C1 are shown in Fig. 4.1b. These are obtained by interpolating the position readouts from two BPMs which will be explained in Chap. 5. The beam was moved in a 2D grid manner. Our task is to find correlations between the HOM spectra and the beam positions.

For the ease of the following analysis, we arrange spectra and beam positions in matrix forms. The matrix A contains the magnitude of HOM spectra (real and positive) and is constructed as

$$A = \begin{pmatrix} spectrum_1 \\ spectrum_2 \\ \vdots \\ spectrum_m \end{pmatrix} = (a_1, a_2, \ldots, a_n) \in \mathbb{R}^{m \times n}. \tag{4.1}$$

P. Zhang, *Beam Diagnostics in Superconducting Accelerating Cavities*, Springer Theses, DOI: 10.1007/978-3-319-00759-5_4,

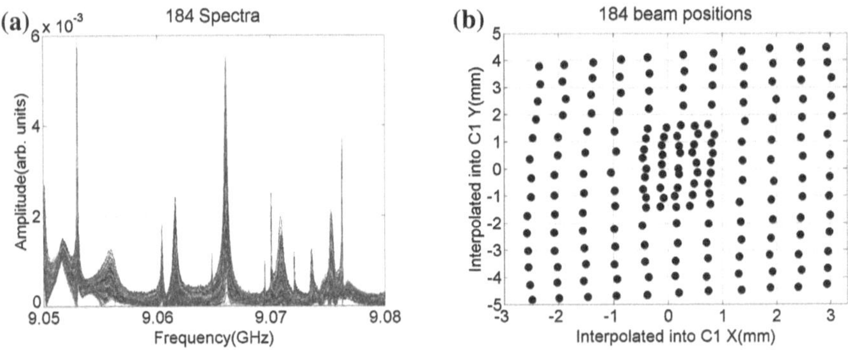

Fig. 4.1 HOM spectra (**a**) measured from coupler C1H1 for 184 transverse beam positions (**b**) at the center of C1

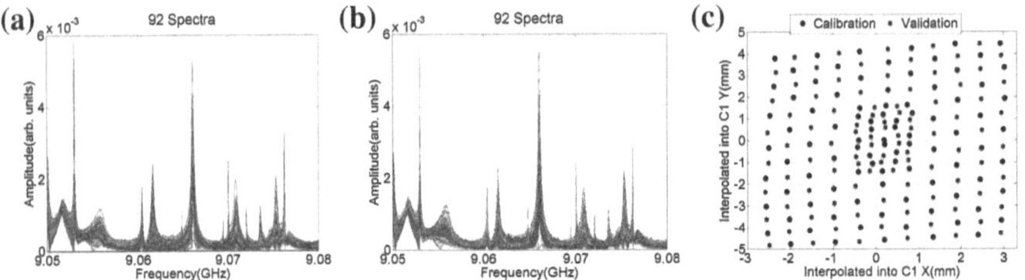

Fig. 4.2 Calibration and validation samples (**a, b**). Spectra are measured from HOM coupler C1H1. The beam positions are interpolated into C1 (**c**) was reprinted with permission from [3]. Copyright 2012, American Institute of Physics

A is a $m \times n$ matrix. $a_i \in \mathbb{R}^m$ is the ith column of A, which contains values of the ith sampling point from all m spectra. In our case, each spectrum has 6241 sampling points, therefore $n = 6241$.

The matrix B contains the beam positions and is constructed as

$$B = \begin{pmatrix} x_1 & y_1 \\ x_2 & y_2 \\ \vdots & \vdots \\ x_m & y_m \end{pmatrix} \in \mathbb{R}^{m \times 2}. \tag{4.2}$$

B is a $m \times 2$ matrix. x_i and y_i are the horizontal and vertical beam coordinates of the ith beam position at the center of C1 corresponding to the measurement of $spectrum_i$.

We have built a system consisting of spectra A and beam positions B. To study the correlations, we split the samples equally into calibrations and validations in order to obtain a full coverage in 2D space for both ones. The spectra are denoted as A_{exp}^{cal} and A_{exp}^{val} and are shown in Figs. 4.3a, b. The corresponding beam positions are shown in

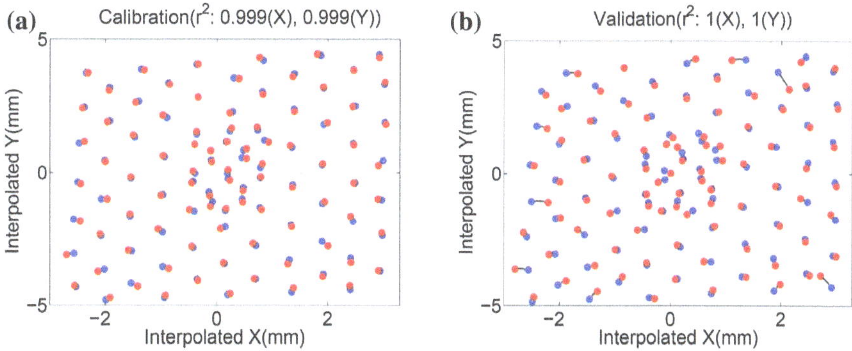

Fig. 4.3 Measurements (*blue*) and predictions (*red*) of transverse beam positions from calibration and validation samples. The applied method is DLR. Points connected with black lines belong to the same beam position. Reprinted with permission from [3]. Copyright 2012, American Institute of Physics

Fig. 4.3c and denoted as B^{cal}_{exp} and B^{val}_{exp}. Therefore, both calibration and validation samples have $m = 92$ in Eqs. 4.1 and 4.2. In the following sections, calibration samples will be used to calculate the calibration coefficients, which relate HOM spectra to beam positions. Then these coefficients are applied on the HOM spectra from validation samples to make predictions of the beam positions. The differences between predicted positions and measured ones are used to evaluate the performance of various methods.

4.2 Direct Linear Regression

As discussed in Chap. 1, dipole modes have linear correlations to transverse beam offsets. For the extraction of beam position information from HOM spectra, a straightforward method is *Direct Linear Regression* (DLR),

$$A \cdot M = B, \tag{4.3}$$

where $A \in \mathbb{R}^{m\times(n+1)}$ with one column of 1 added to the spectra matrix defined in Eq. 4.1 representing the intercept term, $B \in \mathbb{R}^{m\times 2}$ is the position matrix defined in Eq. 4.2, $M \in \mathbb{R}^{(n+1)\times 2}$ is the matrix of regression coefficients. M is further represented as $\left(M_x, M_y\right)$, where each is a $(n+1)\times 1$ vector. The matrix M is obtained by solving the linear system composed by matrices A and B [2]. The algorithm used to solve the linear system is least squares and explained in Appendix A.1.

The calibration samples A^{cal}_{exp} and B^{cal}_{exp} are used to determine the matrix M, then the predictions of B can be obtained by

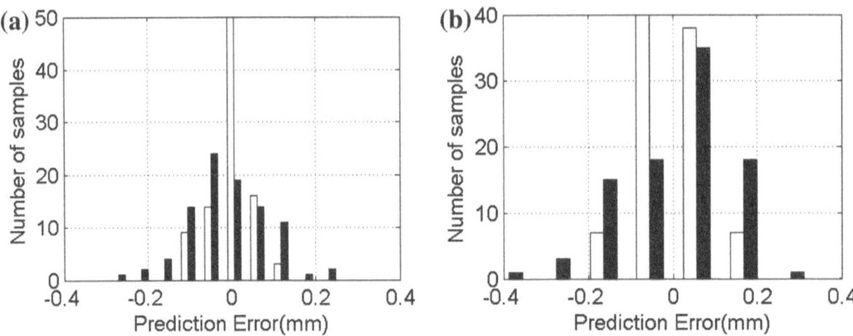

Fig. 4.4 Prediction errors of the beam position for calibration (*white*) and validation samples (*black*). The applied method is DLR. Reprinted with permission from [3]. Copyright 2012, American Institute of Physics

$$B_{pre}^{cal} = A_{exp}^{cal} \cdot M, \tag{4.4a}$$

$$B_{pre}^{val} = A_{exp}^{val} \cdot M, \tag{4.4b}$$

where B_{pre}^{cal} and B_{pre}^{val} are predictions of B for calibration and validation samples respectively. Figure 4.3a shows the measured beam positions B_{exp}^{cal} along with the predicted beam positions B_{pre}^{cal} for calibration samples. They almost overlap, which proves the linear dependence of HOM spectra on transverse beam offsets. The linearity is further validated in Fig. 4.3b when using the regression coefficients M on the validation samples by Eq. 4.4b. The coefficient of determination r^2 is used to measure the consistency and is shown in the figure ($r^2 = 1$, perfect fit; $r^2 = 0$, poor fit; see Appendix A.3).

The prediction errors ΔB^{cal} and ΔB^{val} are also calculated as

$$\Delta B^{cal} = B_{exp}^{cal} - B_{pre}^{cal}, \tag{4.5a}$$

$$\Delta B^{val} = B_{exp}^{val} - B_{pre}^{val}. \tag{4.5b}$$

Figure 4.4 shows the histograms of prediction errors for calibration and validation samples by using DLR. The rms of the prediction error (rms error) is defined as

$$E_{rms}^{cal} = \sqrt{\frac{1}{m}\sum_{i=1}^{m}\left(\Delta B_i^{cal}\right)^2}, \tag{4.6a}$$

$$E_{rms}^{val} = \sqrt{\frac{1}{m}\sum_{i=1}^{m}\left(\Delta B_i^{val}\right)^2}. \tag{4.6b}$$

Table 4.1 RMS of prediction errors using DLR

DLR	x (mm)	y (mm)
E^{cal}_{rms}	0.05	0.08
E^{val}_{rms}	0.10	0.15

E^{cal}_{rms} and E^{val}_{rms} have the unit of millimeter, the same as B^{cal}_{exp}, B^{val}_{exp}, B^{cal}_{pre} and B^{val}_{pre}. Table 4.1 lists E^{cal}_{rms} and E^{val}_{rms} obtained by using DLR. The rms errors for calibration and validation samples are comparably small, which means the calibration is robust.

The algorithm used in solving the linear system (in the form of Eq. 4.3) is least squares, which is a popular method to solve linear regression problems. The details of this algorithm are explained in Appendix A.1. The method relies on minimizing the difference of predictions (B^{cal}_{pre}) and measurements (B^{cal}_{exp}) whilst modifying the elements of the matrix M. In our case, the size of matrix M is 6242 × 2, which is directly related to the number of sampling points present in the spectra matrix A. In this case, a considerable number of unknown variables needs to be determined. This is computationally expensive. As all sampling points are used in the regression, each column of the spectra matrix A, a_i, is a regressor. The correlation coefficients $R(i, j)$ of these regressors are calculated by [4]

$$R(i, j) = \frac{\mathrm{Cov}(a_i, a_j)}{\sqrt{\mathrm{Cov}(a_i, a_i) \cdot \mathrm{Cov}(a_j, a_j)}}, \tag{4.7}$$

where a_i and a_j are ith and jth column of the spectra matrix A as defined in Eq. 4.1, $\mathrm{Cov}(a_i, a_j)$ is the covariance [5] between two regressors a_i and a_j. In this definition, $R = \pm 1$ means strong correlations while $R = 0$ means no correlations. Figure 4.5 shows the correlation coefficients of calibration samples. Most regressors are strongly correlated and this makes the linear system sensitive to fluctuations of the sampling points' values. In our case, it means that the system is vulnerable to noise. Moreover,

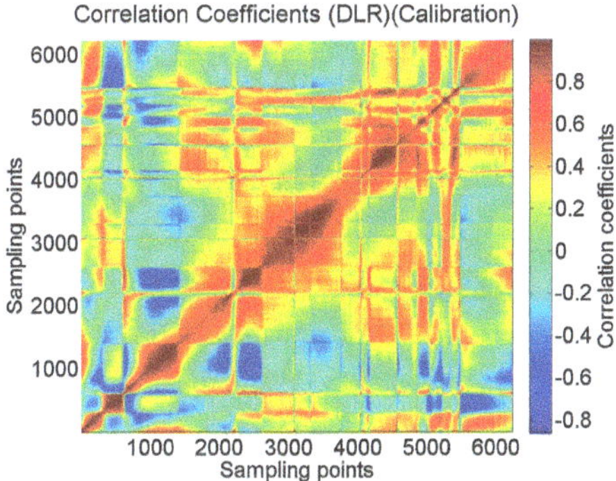

Fig. 4.5 Correlation coefficients between each regressor of DLR. Strong correlations are denoted in *red* and *blue*

even though it is not obvious here, overfitting is always a potential problem unless the number of calibration samples is fairly large ($m > n$). To avoid these risks, it is advisory to reduce the system size from multi-thousand to several tens, and then apply linear regression on the reduced system. The methods we used for this purpose are described in the following sections.

4.3 Singular Value Decomposition

In order to reduce the dimension of the system, a method known as *Singular Value Decomposition* (SVD) is used to find a small number of prominent components from the spectra matrix A. In general, SVD looks for the patterns of a matrix in terms of SVD modes without the knowledge of any physical parameters (like dipole mode frequency, quality factor, beam position, etc.). Those SVD modes are natural groupings within the matrix, which are not clearly visible or explicitly defined in the matrix itself.

Applying SVD, the matrix A is decomposed into the product of three matrices (see Appendix A.2),

$$A = U \cdot S \cdot V^T, \tag{4.8}$$

where V^T indicates the transpose of V. The columns of U and V are singular vectors of A, which form the bases of the decomposition. S is a diagonal matrix whose non-zero elements are known as *singular values*. Applying SVD on calibration samples A_{exp}^{cal} without the intercept term (blue dots in Fig. 4.2c), the singular value of each SVD mode is obtained and shown in Fig. 4.6a. It can be seen that the first few SVD modes have relatively large singular values, in other words, they are the dominant

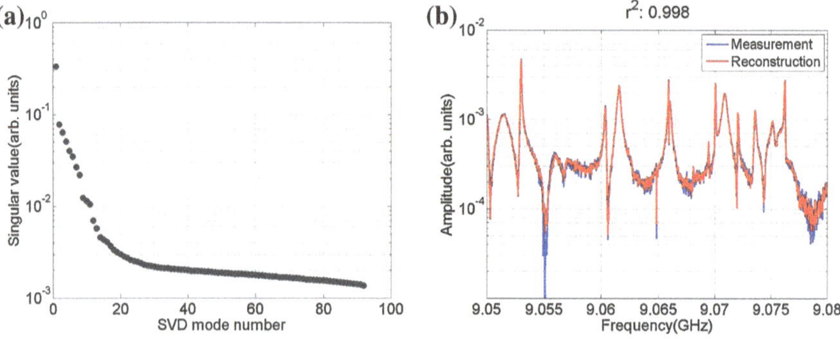

Fig. 4.6 **a** Singular value of each SVD mode. **b** Reconstructed spectrum (*red*) from the first 17 SVD modes compared with the original spectrum (*blue*) for one beam position. Reprinted with permission from [3]. Copyright 2012, American Institute of Physics

patterns or natural groupings of matrix A^{cal}_{exp}. In this case, the first 17 SVD modes are selected for the following analysis.

Each of those SVD modes can be used to produce a spectra matrix A^{cal}_i by

$$A^{cal}_i = U_i \cdot S_{ii} \cdot V^T_i, \tag{4.9}$$

where U_i is the ith column of U, and V^T_i is the ith row of V^T. S_{ii} denotes the ith diagonal element of S. A^{cal}_i has the same size as A^{cal}_{exp}. Figure 4.7 shows one spectrum for each of the first nine A^{cal}_i. The approximation A^{cal}_{reco} can be calculated by simply summing over the related A^{cal}_i's as

$$A^{cal}_{reco} = \sum_{i=1}^{p} A^{cal}_i. \tag{4.10}$$

A^{cal}_{reco} is plotted together with A^{cal}_{exp} in Fig. 4.6b for $p = 17$. It can be seen that the spectrum is well represented by using only the first 17 SVD modes, which is also illustrated by $r^2 \approx 1$. The choice of p will be discussed later.

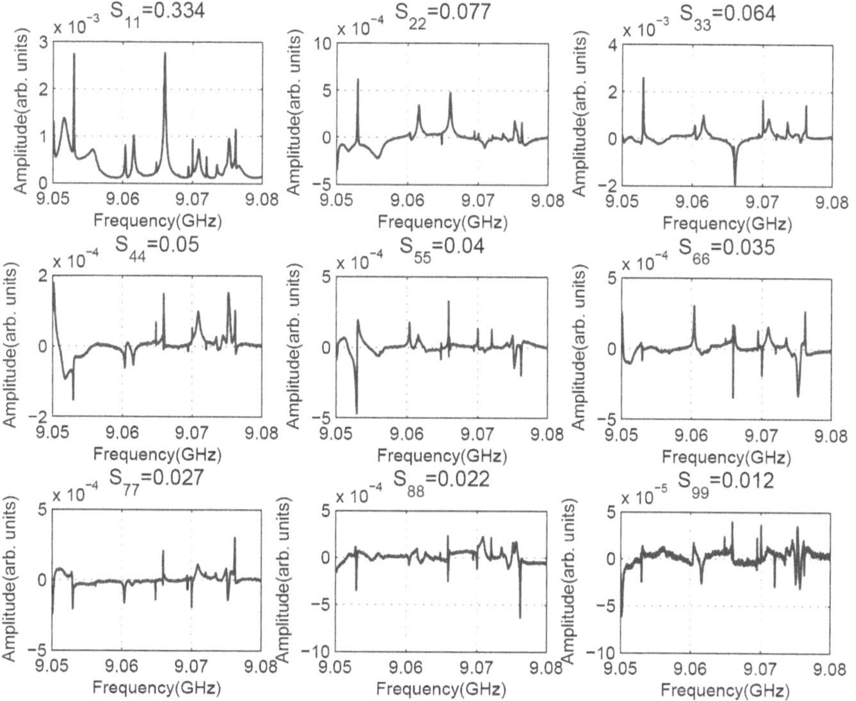

Fig. 4.7 Spectra reconstructed from each of the first nine SVD modes. The first six SVD modes plots were reprinted with permission from [3]. Copyright 2012, American Institute of Physics

Having the SVD base vectors from the decompositions of spectra matrix A^{cal}_{exp}, the amplitude of all SVD modes, $A^{cal}_{svd,full}$, can be obtained for each beam position by

$$A^{cal}_{svd,full} = A^{cal}_{exp} \cdot V = (A^{svd}_1, A^{svd}_2, \cdots). \tag{4.11}$$

$A^{svd}_i \in \mathbb{R}^{m \times 1}$ contains the mode amplitude for each beam position of the ith SVD mode. Figure 4.8a shows the mode amplitude of all 92 SVD modes with non-zero singular values obtained from the calibration samples. The correlation coefficient between each pair of SVD mode is also calculated using Eq. 4.7 by replacing a_i and a_j with A^{svd}_i and A^{svd}_j and it is shown in Fig. 4.8b. The first SVD mode has relatively strong correlations with several SVD modes. This is because the mean of the 92 calibration spectra is not zero (Fig. 4.9).

Along with a column of one, $I \in \mathbb{R}^{m \times 1}$, representing the intercept term, we use the first p ($p < m$) SVD modes to build the SVD mode amplitude matrix A^{cal}_{svd}:

$$A^{cal}_{svd} = (I, A^{svd}_1, A^{svd}_2, \cdots, A^{svd}_p), \tag{4.12}$$

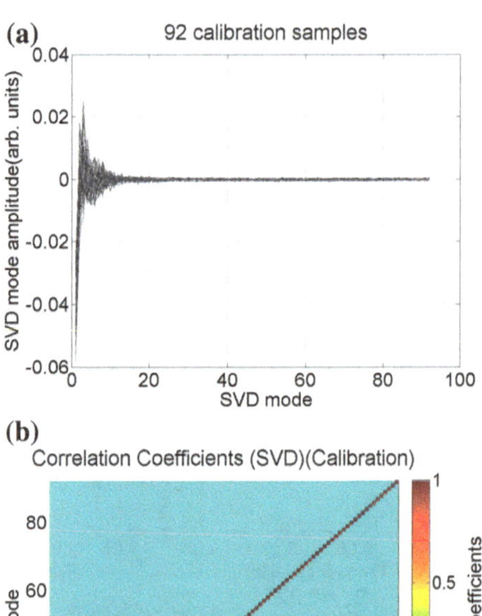

Fig. 4.8 (a) SVD mode amplitudes. (b) SVD mode correlations

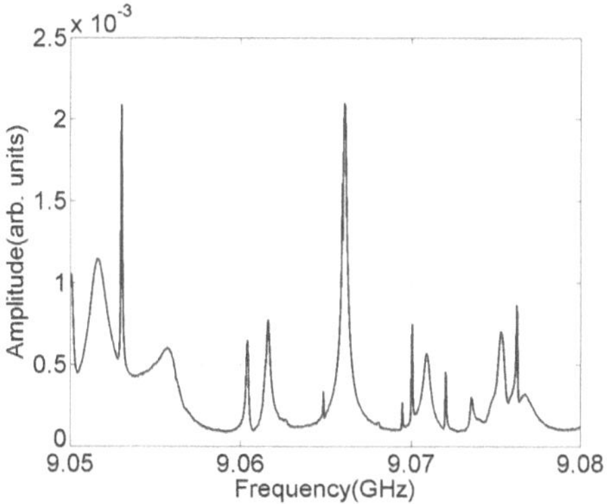

Fig. 4.9 Mean spectrum of the 92 calibration samples

The size of the matrix A^{cal}_{svd} is significantly smaller ($p + 1$ columns, representing p SVD modes and 1 intercept term) compared to the original spectra matrix A^{cal}_{exp} (6242 columns, representing 6241 sampling points and 1 intercept term). Replacing A^{cal}_{exp} by A^{cal}_{svd} in the regression (Eq. 4.3), the linear system composed by A^{cal}_{svd} and B^{cal}_{exp} is now over-determined, and has a best solution in a least square sense. The position prediction for calibration samples can be obtained by replacing A^{cal}_{exp} with A^{cal}_{svd} in Eq. 4.4a. The prediction for validation samples has two steps: first, project A^{val}_{exp} onto the base vectors obtained from calibration samples to get the amplitude of SVD modes by Eq. 4.11 and construct the SVD mode amplitude matrix A^{val}_{svd} by using Eq. 4.12; second, predict positions using Eq. 4.4b by replacing A^{val}_{exp} with A^{val}_{svd}.

The contribution of using the first p SVD modes to determine the transverse beam position x and y is measured by r^2 and $\bar{r}^2$ as shown in Fig. 4.10. $\bar{r}^2$ is the adjusted r^2 calculated as (Appendix A.3)

$$\bar{r}^2 = 1 - \left(1 - r^2\right) \frac{N - 1}{N - p - 1}, \tag{4.13}$$

where N is the total number of SVD modes decomposed from A^{cal}_{exp}, p is the number of SVD modes used in the regression. Compared to r^2, $\bar{r}^2$ measures the net gain by adding new SVD modes into the regression. The rms errors for calibration and validation samples are also calculated by using the first p SVD modes as shown in Fig. 4.11. One can see that using the first 17 SVD modes is seen to give an optimal performance for the x plane. It is also a reasonable choice for the y plane even though less SVD modes would be sufficient. This confirms, with Fig. 4.6, that one can get the majority of beam position information by using the first few SVD modes.

Figure 4.12 shows the HOM response from calibration and validation samples by using the first 17 SVD modes to determine the transverse beam position x and y.

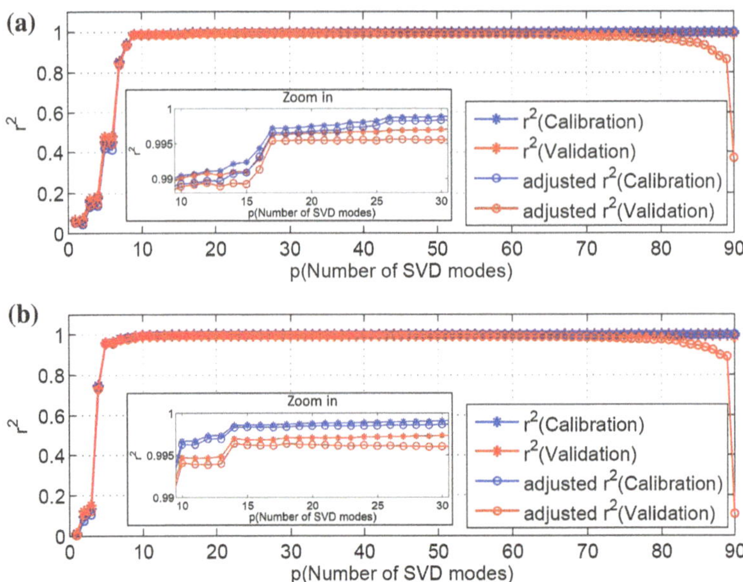

Fig. 4.10 Contribution of the first p SVD modes to determine the transverse beam position x and y measured by r^2 and $\bar{r}^2$

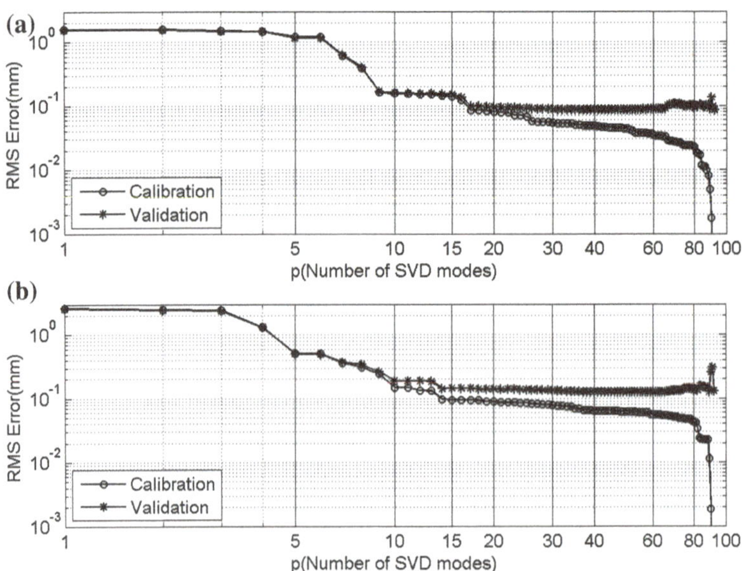

Fig. 4.11 Contribution of the first p SVD modes to determine the transverse beam position x and y measured by the rms error. Reprinted with permission from [3]. Copyright 2012, American Institute of Physics

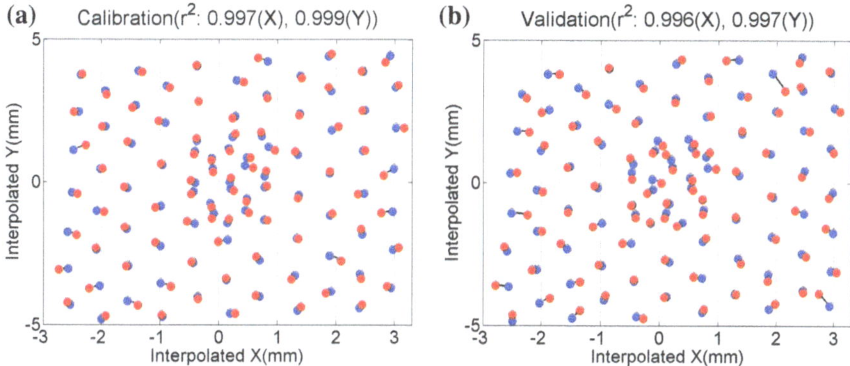

Fig. 4.12 Measurements (*blue*) and predictions (*red*) of the transverse beam position from calibration and validation samples. The applied method is SVD with the first 17 SVD modes. Points connected with *black lines* belong to the same beam position. Reprinted with permission from [3]. Copyright 2012, American Institute of Physics

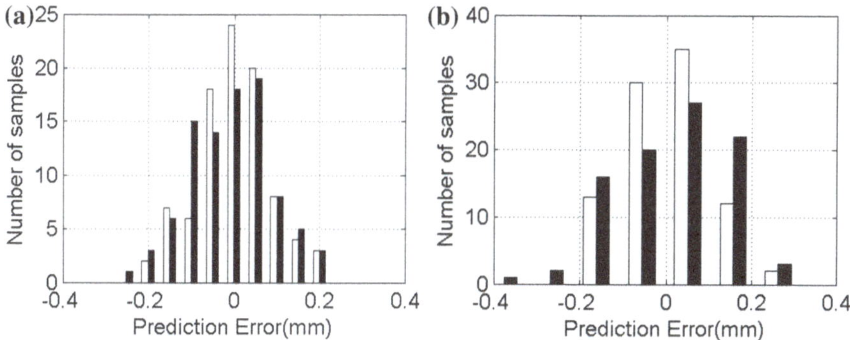

Fig. 4.13 Prediction errors of the beam position for calibration (*white*) and validation samples (*black*). The applied method is SVD with the first 17 SVD modes. Reprinted with permission from [3]. Copyright 2012, American Institute of Physics

The rms errors remain small and comparable for calibration and validation samples as shown in Fig. 4.13 and Table 4.2.

Being a row-based method, SVD is able to effectively cut the signal size down by two orders of magnitude while retaining the concealed beam position information. In the next section, another dimension reduction method namely k-means clustering is explained. Unlike the SVD, k-means clustering is a column-based method.

Table 4.2 RMS of prediction errors using SVD with the first 17 SVD modes.

SVD	x (mm)	y (mm)
E_{rms}^{cal}	0.08	0.09
E_{rms}^{val}	0.10	0.14

4.4 k-means Clustering

As shown in Fig. 4.5, an information redundancy is present in the linear system composed of the spectra matrix A (Eq. 4.1) and the beam position matrix B (Eq. 4.2), as the regressors a_i's from the spectra matrix A are highly correlated. A technique named *k-means clustering* [6] is used in order to group the data into a few cohesive *clusters*. Each $a_i \in \mathbb{R}^m$ is considered as one point in the m-dimensional space. The goal is to find k groups of points based on Euclidean distance. The cluster centroid μ_j is defined as

$$\mu_j = \frac{1}{N} \sum_{i=j_1}^{j_N} a_i, \tag{4.14}$$

where $(j_1, j_2, \cdots, j_N)$ are N indices of the m-dimensional points a_i's which belong to the jth cluster whose centroid is μ_j. Figure 4.14 shows 26 clusters (red) partitioned from calibration samples (blue) for one beam position. The values of the corresponding 26 centroids are shown in Fig. 4.15a for all 92 beam positions. The correlation coefficients of these clusters can be calculated by replacing a_i with μ_j in Eq. 4.7 and are shown in Fig. 4.15b. As expected, correlations can be observed among clusters.

The contribution of different number of clusters to determine the transverse beam position x and y is measured by the rms error shown in Fig. 4.16. Using 26 clusters is seen to give an optimal performance for both x and y.

Figure 4.17 shows the HOM response from calibration and validation samples by using 26 clusters to determine the transverse beam position x and y respectively. The rms errors remain small and comparable for calibration and validation samples as shown in Fig. 4.18 and Table 4.4.

4.5 Comparison of DLR, SVD and k-means Clustering

The performance of the three methods used to extract beam positions described in the previous section is evaluated by two different techniques: the fixed sample split and cross-validation. A figure of merit, RMS of the prediction error (Eq. 4.6) is calculated.

4.5.1 Fixed Sample Split

For the fixed sample split shown in Fig. 4.2c, the measured beam positions are shown in Fig. 4.19 along with predicted positions by the three different methods. The rms errors are listed in Table 4.4. The results obtained are comparable.

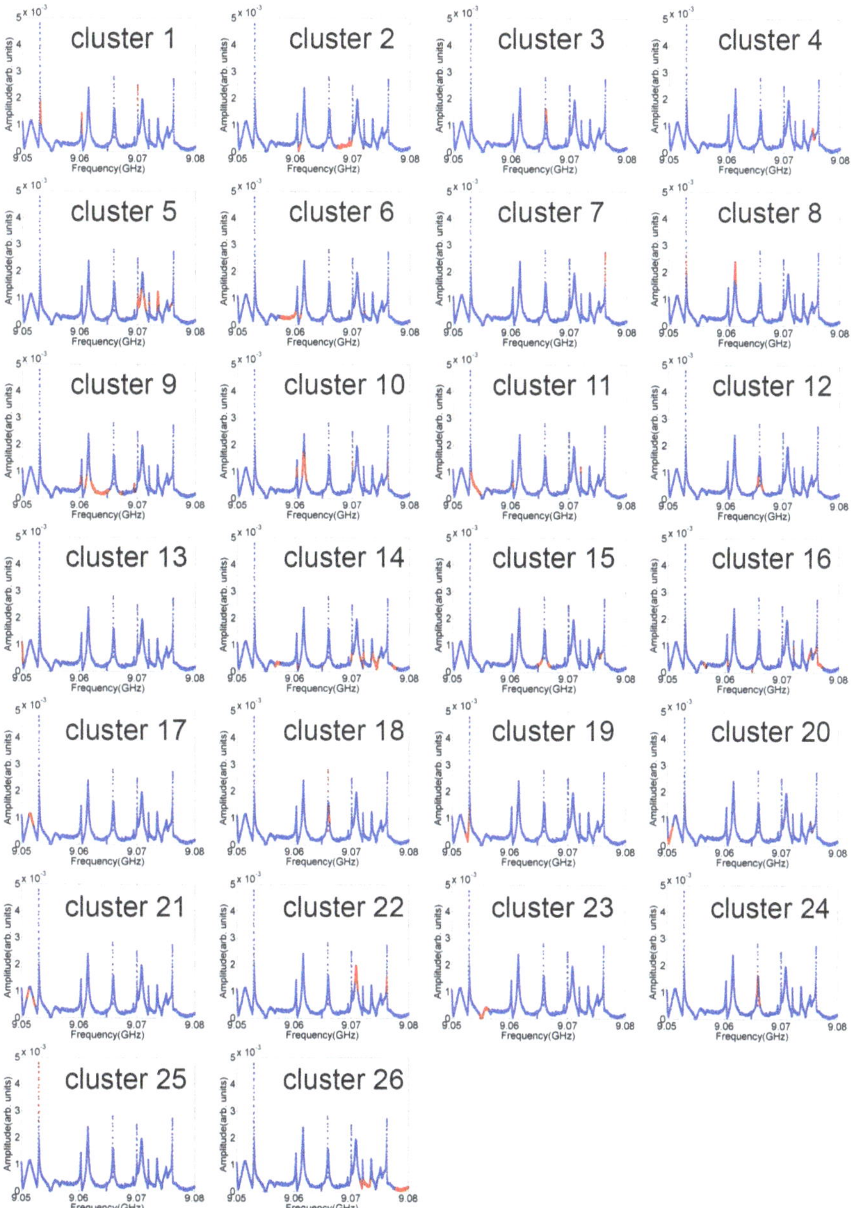

Fig. 4.14 Clusters (*red*) partitioned from calibration samples (*blue*) for one beam position

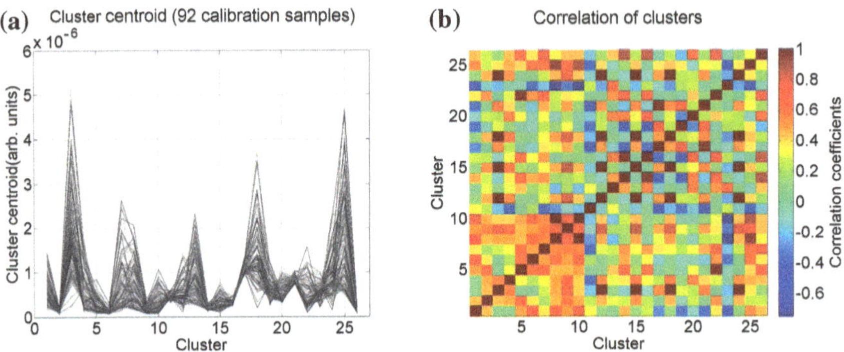

Fig. 4.15 Cluster centroids and their correlations

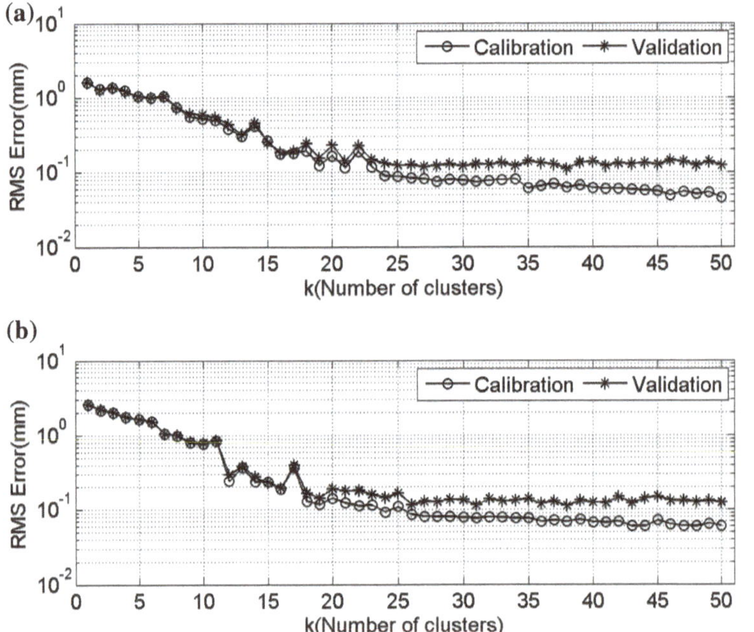

Fig. 4.16 Contribution of k clusters to determine the transverse beam position x and y measured by the rms error

4.5.2 Cross-Validation

Until now, the analysis is based on a specific sample split (Fig. 4.2c). To study the dependence on how samples are split, a technique named cross-validation is used [7]. The 184 samples are split randomly into 129 calibration samples (approximately 70 % of the total) and 55 validation samples (the remaining 30 %). Figure 4.20 shows 12 different random splits.

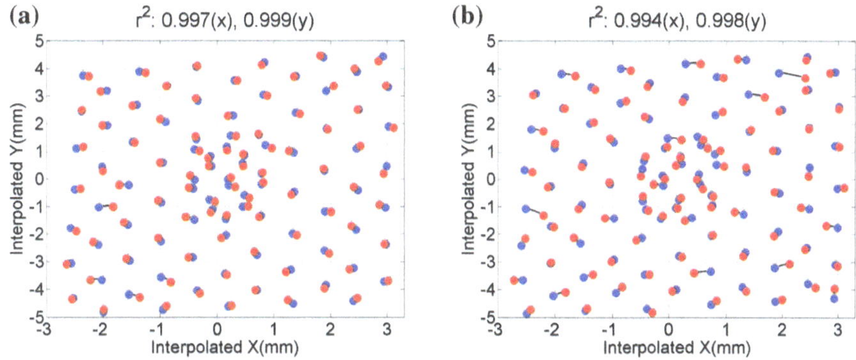

Fig. 4.17 Measurements (*blue*) and predictions (*red*) of the transverse beam position from calibration and validation samples. The applied method is k-means clustering with 26 clusters. Points connected with black lines belong to the same beam position

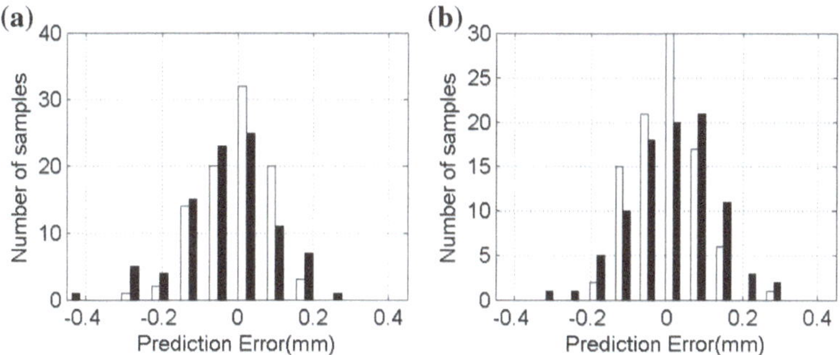

Fig. 4.18 Prediction errors of the beam position for calibration (*white*) and validation samples (*black*). The applied method is k-means clustering with 26 clusters

Table 4.3 Direct comparison of DLR, SVD and k-means clustering for the fixed sample split (Fig. 4.2c). The signal is measured from HOM coupler C1H1

9.05–9.08 GHz	E_{rms}^{cal} (x) mm	E_{rms}^{cal} (y) mm	E_{rms}^{val} (x) mm	E_{rms}^{val} (y) mm
DLR	0.05	0.08	0.10	0.15
SVD	0.08	0.09	0.10	0.14
Clustering	0.08	0.09	0.13	0.12

Table 4.4 RMS of prediction errors using k-means clustering with 26 clusters

Clustering	x(mm)	y(mm)
E_{rms}^{cal}	0.08	0.09
E_{rms}^{val}	0.13	0.12

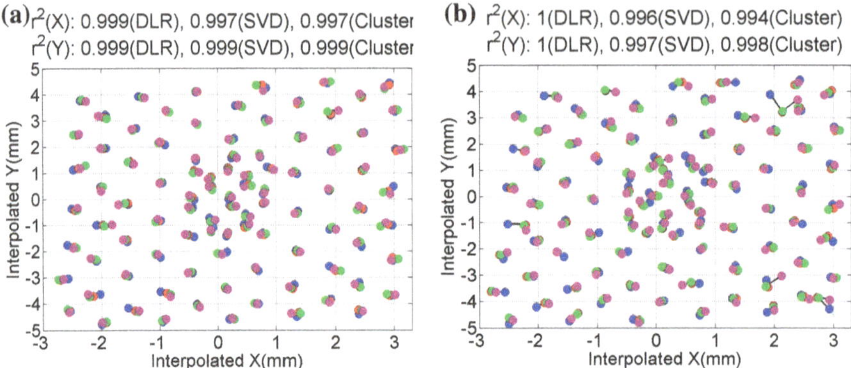

Fig. 4.19 Measurements (*blue*) and predictions of the transverse beam position from calibration and validation samples. The DLR is in *red*. The SVD is in *green* with the first 17 SVD modes. The *k*-means clustering is in magenta with 26 clusters. Points connected with *black lines* belong to the same beam position

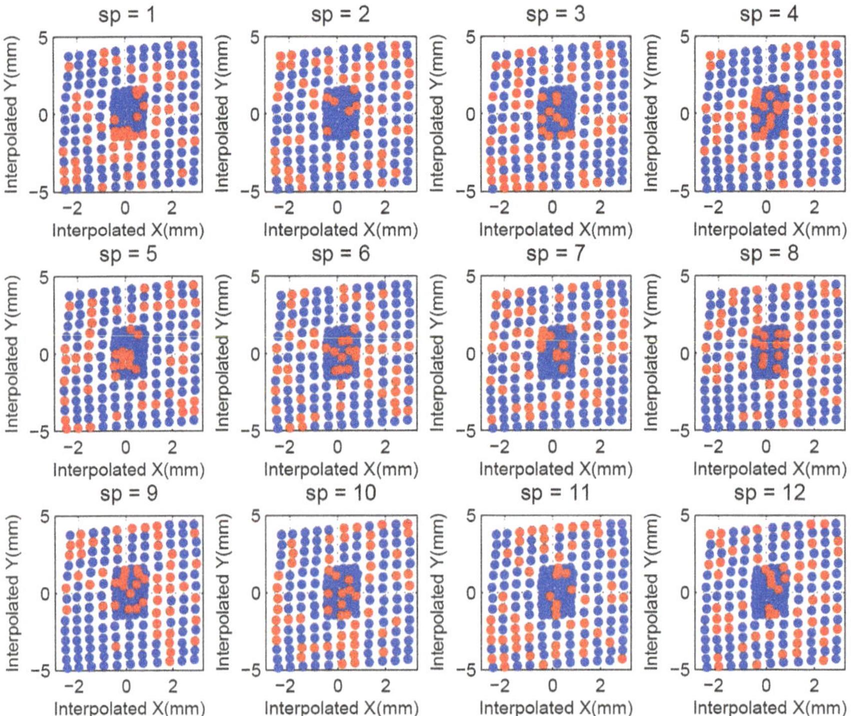

Fig. 4.20 Random splits of the total samples into calibrations (*blue*) and validations (*red*). *sp* numbers the split. The first 8 plots were reprinted with permission from [3]. Copyright 2012, American Institute of Physics

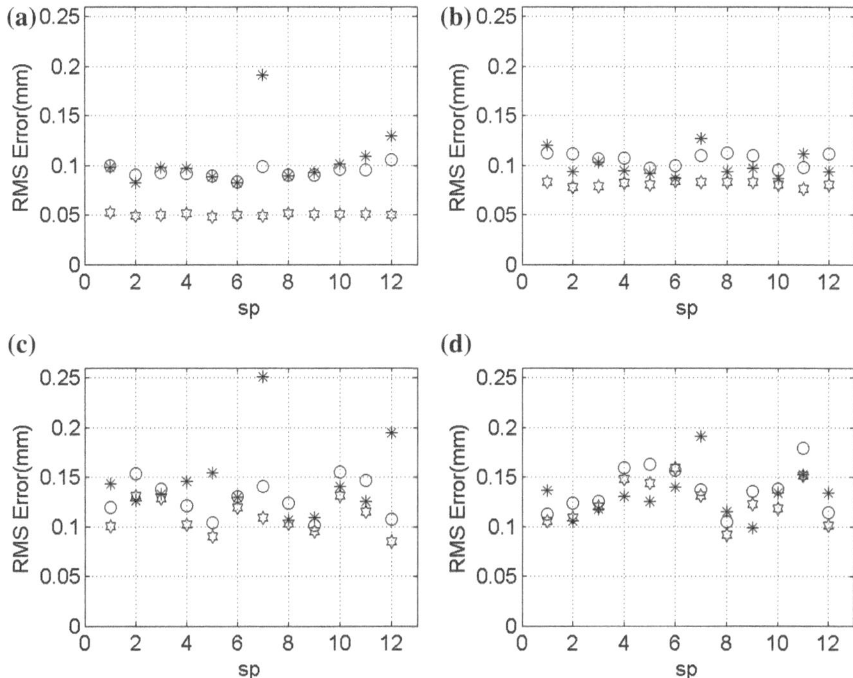

Fig. 4.21 RMS errors of calibration and validation samples from 12 different splits (Fig. 4.20) by using DLR (hexagrams), SVD (*circles*) with the first 17 SVD modes and k-means clustering (*asterisks*) with 26 clusters

Applying DLR, SVD and k-means clustering on the 12 different splits, the responses of HOM spectra to the transverse beam positions represented by rms errors are shown in Fig. 4.21. The variation of the rms errors from most sample splits remains small, which indicates that the HOM response does not rely on a specific sample split. Most of the rms errors stay below 0.15 mm for both x and y planes. However, one notices that the rms error of x is very large for the 7th sample split by using k-means clustering, and this is because clustering relies on a good coverage in 2D space in order to characterize correctly the correlations among m-dimensional points, and consequently determine the correct clusters. Thus, compared to SVD and DLR, k-means clustering is fundamentally more sensitive to the sample split. From Fig. 4.21, the performance of three methods are rather comparable, but SVD and k-means clustering is far superior to DLR in terms of the computation time of the coefficient matrix M.

Equipped with three methods for beam information extraction, the dependencies of dipole modes on beam offsets will be studied in Chap. 5. These methods are also used in the analysis of HOM-based beam diagnostics to be presented in Chap. 6.

References

1. S. Walston and others, Performance of a high resolution cavity beam position monitor system. Nucl. Instrum. Methods Phys. Res. A **578**(1), 1–22 (2007)
2. P. Zhang, N. Baboi, R.M. Jones, Statistical methods for transverse beam position diagnostics with higher order modes in third harmonic 3.9 GHz superconducting accelerating cavities at FLASH. Nucl. Instrum. Methods Phys. Res. A (2012). http://dx.doi.org/10.1016/j.nima.2012.11.057
3. P. Zhang, N. Baboi, R.M. Jones, I.R.R. Shinton, T. Flisgen, H.W. Glock, A study of beam position diagnostics using beam-excited dipole modes in third harmonic superconducting accelerating cavities at a free-electron laser. Rev. Sci. Instrum. **83**, 085117 (2012)
4. R.L. Ott, M. Longnecker, *An Introduction to Statistical Methods and Data Analysis*, ch. 11.7, 6th edn. (Duxbury Press, Belmont, 2008), p. 609
5. M.R. Spiegel, J.J. Schiller, R.A. Srinivasan, *Schaum's Outline of Theory and Problems of Probability and Statistics*, ch. 8, 2nd edn. (McGraw-Hill, New York, 2000), p. 281
6. C.M. Bishop, *Pattern Recognition and Machine Learning*, ch. 9.1, 1st edn. (Springer Publishing, New York, 2006), pp. 424–428
7. R. Kohavi, A study of cross-validation and bootstrap for accuracy estimation and model selection, in *Proceedings of the 14th International Joint Conference on Artificial Intelligence*, vol. 2 (Montreal, Quebec, 1995), pp. 1137–1143

Chapter 5
Dependencies of HOM on Transverse Beam Offsets

After characterizing the modes in the coupled third harmonic cavities in ACC39, experiments using the RSA were conducted to study the correlations of dipole modes to transverse beam positions. The goal is to search suitable dipole modes for position diagnostics [1]. A superb position resolution is not the focus of this study as the RSA has only limited resolution [2], which can be improved by dedicated electronics. Localized modes and coupled modes are studied with various data analysis techniques: Lorentzian fit, DLR, SVD and k-means clustering. The latter three methods have been introduced in Chap. 4, and their prediction accuracies are compared in this chapter for both localized and coupled modes.

5.1 Measurement Scheme

In order to investigate the dependence of dipole modes on transverse beam positions, one needs to create various transverse beam offsets in the cavity. The schematic of the measurement setup is shown in Fig. 5.1. An electron bunch of approximately 0.5 nC was accelerated on-crest through ACC1 to approximately 150 MeV before entering the ACC39 module. Two steering magnets located upstream of ACC1 were used to produce horizontal and vertical offsets of the electron bunch in ACC39. Two beam position monitors (BPM-A and BPM-B) were used to record transverse beam positions before and after ACC39. By switching off the accelerating field in ACC39 and all nearby quadrupoles, a straight line trajectory of the electron bunch was produced between those two BPMs. Therefore, the transverse offset of the electron bunch in each cavity can be determined by interpolating the readouts of the two BPMs (see Fig. C.1).

For each beam position, along with HOM signals, nearby toroids, BPMs and currents of steering magnets were also recorded synchronously. The electron bunch was moved in two-dimensional (2D) cross manner and 2D grid manner. The steerers induced changes in beam offset and trajectory angle in ACC39. The contribution of

P. Zhang, *Beam Diagnostics in Superconducting Accelerating Cavities*, Springer Theses, DOI: 10.1007/978-3-319-00759-5_5,

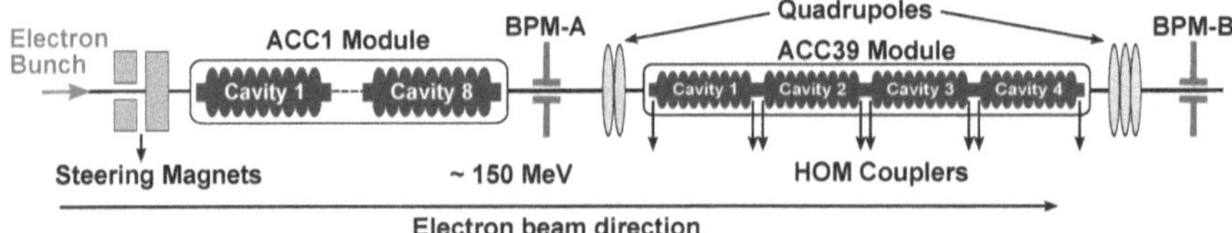

Fig. 5.1 Schematic of measurement setup for HOM dependence study (not to scale, cavities in ACC1 are approximately three times larger than those in ACC39). Reprinted with permission from [1]. Copyright 2012, American Institute of Physics

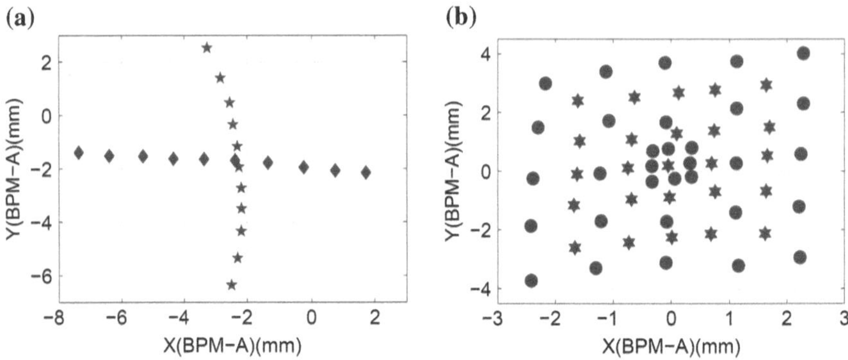

Fig. 5.2 Readings of BPM-A during beam scans. **a** 2D cross-like scan. **b** 2D grid-like scan. Reprinted with permission from [1]. Copyright 2012, American Institute of Physics

the angle wakefield is much smaller than that of the offset wakefield [3], the angle is therefore not considered in this study. Besides, there are also considerable technical difficulties to create trajectory angle in ACC39 independent of offsets. The readings from BPM-A for these two scans are shown in Fig. 5.2. Position interpolations are applicable throughout this chapter except the 2D cross scan (Fig. 5.2a) used for measuring beam-pipe modes. In this case, the quadrupoles between BPM-A and BPM-B were still on. One notices some tilt of the position readings during both scans. This is due to the coupling between x and y plane caused partially by the ACC1 module and partially by BPM-A itself.

5.2 The Localized Dipole Beam-Pipe Modes

From simulations (Sect. 2.3) and measurements (Chap. 3), we know that modes at approximately 4.1 GHz are localized dipole beam-pipe modes. The spectra of two modes at approximately 4.1 GHz measured from C2H2 at ten different horizontal beam positions (diamonds in Fig. 5.2a) are shown in Fig. 5.3. The vertical position reading from BPM-A varied by ±0.24 mm during the horizontal scan. Variations of the mode amplitude with respect to the transverse beam position can be observed. The amplitudes have been normalized to 1 nC of bunch charge. The amplitudes do

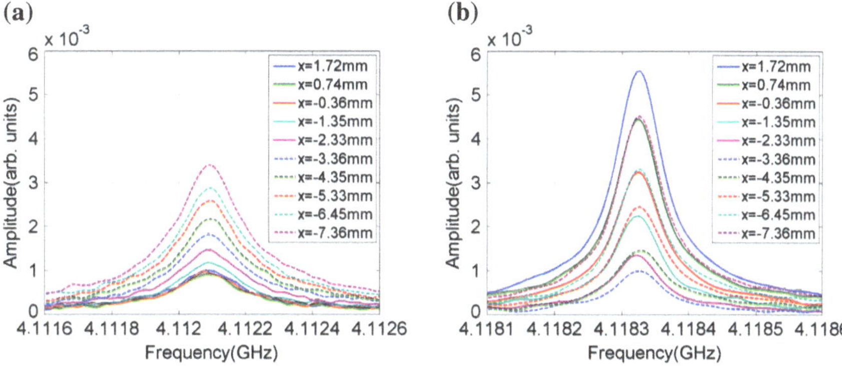

Fig. 5.3 Spectra of dipole beam-pipe modes varies with horizontal beam position read from BPM-A. The spectra were measured from HOM coupler C2H2. **a** Mode #1. **b** Mode #2. Reprinted with permission from [1]. Copyright 2012, American Institute of Physics

not reach zero because of an offset in the vertical direction during the horizontal scan. In addition, other sources such as fabrication errors and cell misalignment can also contribute.

The mode amplitude can be obtained by a Lorentzian fit (Eq. 3.1). Figure 5.4 show the fitted amplitude of each mode as a function of the transverse beam position read from BPM-A for horizontal and vertical beam scans respectively (Fig. 5.2a). An apparently linear dependence of the mode amplitude on the transverse beam position can be observed, which indicates a dipole-like behavior.

The polarizations of these two dipole beam-pipe modes are shown in Fig. 5.5 where the amplitudes were measured in the 2D grid scan (Fig. 5.2b). Position interpolations from the two BPM readouts (BPM-A and BPM-B) are applied to get the transverse beam positions in C2. These two modes are polarized perpendicularly to each other, which indicates potential splitting of the mode degeneracy. This is also representative of a dipole-like behavior. The symmetry of the ideal cylindrical structure is broken by HOM couplers installed on the connecting beam pipes. This causes the frequency split of the two polarizations along with the inevitable manufacturing tolerances. There are three beam pipes inter-connecting cavities in ACC39 (see Fig. 1.6), but clear polarizations can only be observed from the beam pipe connecting C2 and C3. The power couplers installed on the other two beam pipes may account for this. Dependencies and polarizations of beam-pipe modes measured from all eight HOM couplers can be found in Ref. [4].

5.3 Trapped Cavity Modes in the Fifth Dipole Band

As explained in Sect. 2.3 and Chap. 3, there are trapped cavity modes in the fifth dipole band. By moving the beam horizontally, variations of the amplitude of one mode at approximately 9.0562 GHz can be clearly seen in Fig. 5.6a. The vertical position in C2

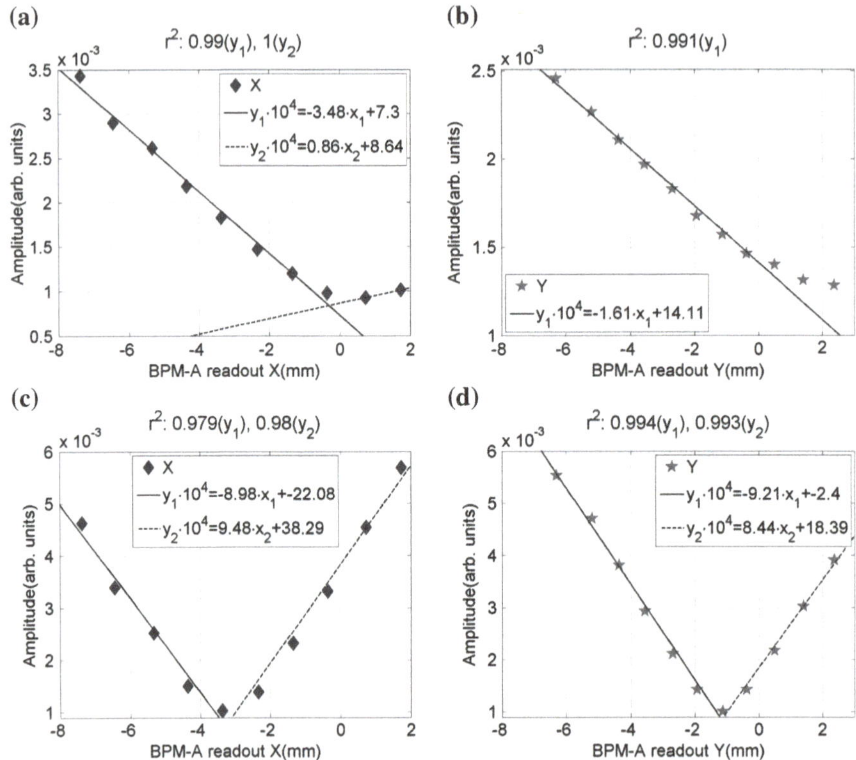

Fig. 5.4 Fitted amplitude of dipole beam-pipe modes as a function of the transverse beam position read from BPM-A. The signals were measured from HOM coupler C2H2. The goodness of linear fit is measured by r^2 (Appendix A.3). **a** Mode #1 (x). **b** Mode #1 (y). **c** Mode #2 (x). **d** Mode #2 (y)

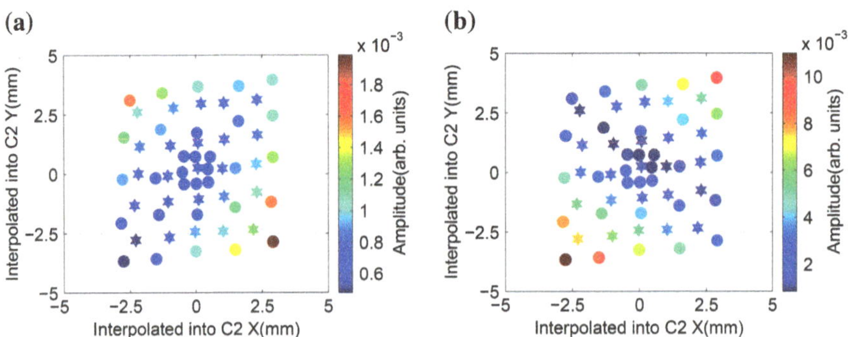

Fig. 5.5 Fitted amplitude of dipole beam-pipe modes as a function of the transverse beam position interpolated into C2. The signals were measured from HOM coupler C2H2. **a** Mode #1. **b** Mode #2. Reprinted with permission from [1]. Copyright 2012, American Institute of Physics

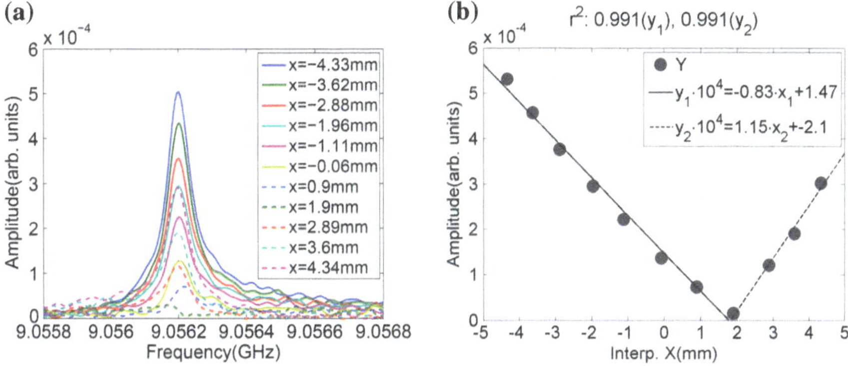

Fig. 5.6 **a** Modal spectra; **b** Fitted amplitude of one dipole mode as a function of the transverse beam position interpolated into C2. The signals were measured from HOM coupler C2H2

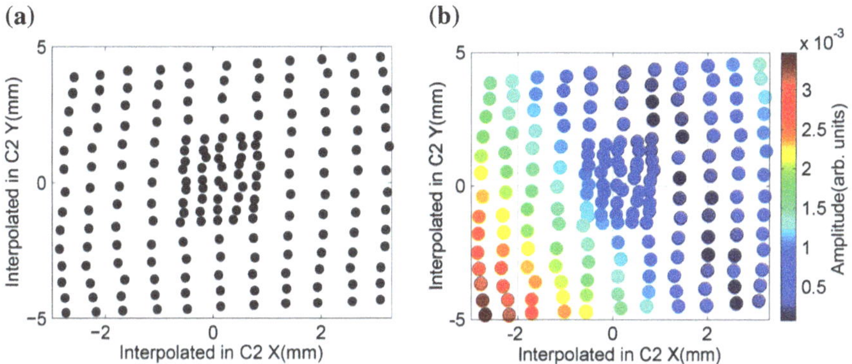

Fig. 5.7 Fitted amplitude of one dipole mode as a function of the transverse beam position interpolated into C2. The signals were measured from HOM coupler C2H2. **a** 2D grid-like scan. **b** Polarization. Reprinted with permission from [1]. Copyright 2012, American Institute of Physics

varied by ± 0.27 mm during this horizontal scan. Using a Lorentzian fit to obtain the amplitudes of the modes and plotting against horizontal beam positions interpolated from BPM readouts, the linear dependence can be clearly seen in Fig. 5.6b, which indicates a dipole-like behavior.

In general, modes in the fifth dipole band have small amplitudes, therefore a dedicated grid-like beam scan was conducted without the 10 dB external attenuators which were used for other frequency bands. Indeed the R/Q of these modes are small (see Table B.4). The interpolated beam positions at the center of cavity C2 are shown in Fig. 5.7a. The amplitude of the same mode at approximately 9.0562 GHz is again obtained by a Lorentzian fit for each beam position, and plotted in Fig. 5.7b. The polarization of this mode can be observed.

The spectrum ranging from 9.05 to 9.08 GHz is used for beam position determination. Modes in this frequency band are trapped in each cavity. The spectra measured

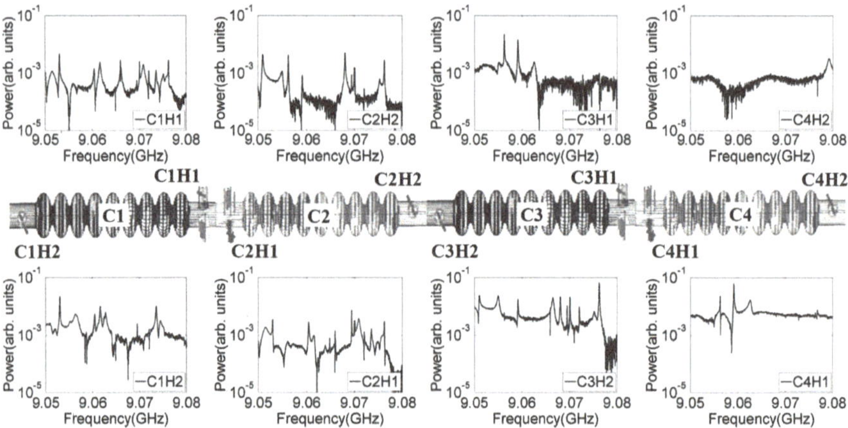

Fig. 5.8 Spectra measured from all eight HOM couplers of ACC39 module at one beam position (the most *top-right point* in Fig. 4.1b). Reprinted with permission from [1]. Copyright 2012, American Institute of Physics

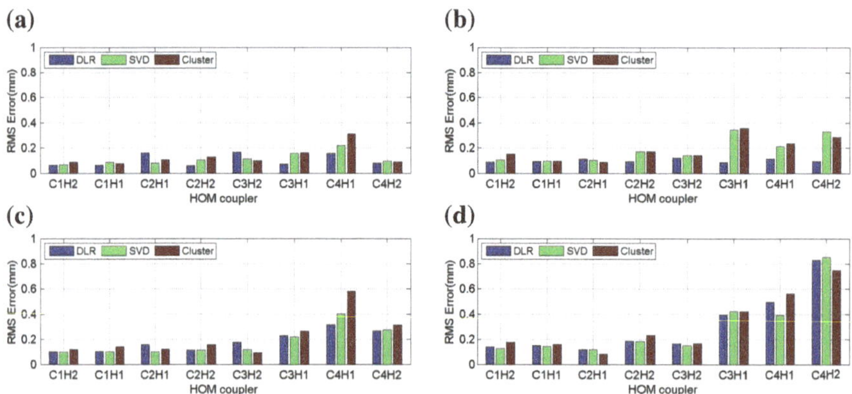

Fig. 5.9 The rms errors of all eight HOM couplers for spectra ranging from 9.05 to 9.08 GHz. The applied method is DLR, SVD with the first 17 SVD modes and k-means clustering with 26 clusters. **a** x (Calibration). **b** y (Calibration). **c** x (Validation). **d** y (Validation)

from all eight HOM couplers are shown in Fig. 5.8 for one beam position. The samples are equally split into 92 calibration samples and 92 validation samples as shown previously in Fig. 4.2c.

DLR, SVD and k-means clustering methods, described in Chap. 4, are used to analyze the data. The rms errors obtained are shown in Fig. 5.9. HOM spectra respond well to the beam movement for most couplers but with individual differences. The cavities are not identical due to fabrication tolerances, which alters the field distributions of the modes. In addition, modes picked up by each coupler vary according to the detailed features of individual couplers. This gives rise to special requirements of using HOMs for beam diagnostics [5], in particular a larger bandwidth is required to accommodate the diverse modal spectra amongst couplers.

Previous work in the TESLA 1.3 GHz cavities [3] used two polarizations of one single dipole mode. We include multiple dipole modes in a narrow frequency band (30 MHz). In our case, it is not possible to separate one specific mode for each coupler, furthermore, multiple modes improve the position prediction accuracies (see Sect. 6.2.4). The number of SVD modes used for regression are determined by the magnitude of singular values (Fig. 4.6a) and the prediction accuracies from validation samples (Fig. 4.11). This turns out to be a larger number (17 SVD modes) than in Ref. [3] (6 SVD modes). This might be due to the fact that modes in the narrow frequency band have individual relationship to the beam movement.

The fact that modes we used are trapped enables the beam position to be determined locally within each cavity. This makes these cavity modes attractive for beam diagnostics compared to the beam-pipe modes, where the EM fields are mainly in the beam pipes and end-cells, and to the first two dipole bands, where the modes couple among cavities (see the following section). However, modes in the fifth dipole band are located in an upper frequency range, and have small coupling to the beam (small R/Q values typically in the range 0.002–2.171 Ω/cm^2). The former will require careful electronics design and the latter will impact position resolution of the electronics (Chap. 6).

5.4 Coupled Cavity Modes in the First and the Second Dipole Band

The electromagnetic field corresponding to most dipole modes are able to travel from cavity to cavity (see Fig. 3.10). Unlike the localized beam-pipe modes and the trapped cavity modes, these multi-cavity modes enable the transverse beam position to be determined for an entire four-cavity module rather than local position inside each cavity or beam pipe. However, the 3.9 GHz four-cavity string is not significantly longer than one TESLA 1.3 GHz cavity (approximately 1.3 times), and hence the position determined based on the four-cavity module is still useful. Moreover, position diagnostics using these modes has the potential for better precision and is relevant for beam alignment due to their strong coupling to the beam (high R/Q values typically in the range 10.572–50.307 Ω/cm^2 for the first dipole band and 5.057–20.877 Ω/cm^2 for the second dipole band) [6].

Simulations showed that the frequency range from 4.9 to 4.95 GHz covers some strong coupling modes, therefore, the spectra within this 50 MHz bandwidth are used for the following analysis. The second dipole band spanning from 5.1 to 5.5 GHz has been studied in Ref. [7]). The beam was moved in a grid manner as shown in Fig. 5.5b. The positions read from BPM-A and BPM-B are interpolated into the center of the module. The samples are split into 32 calibrations and 25 validations as shown in Fig. 5.10.

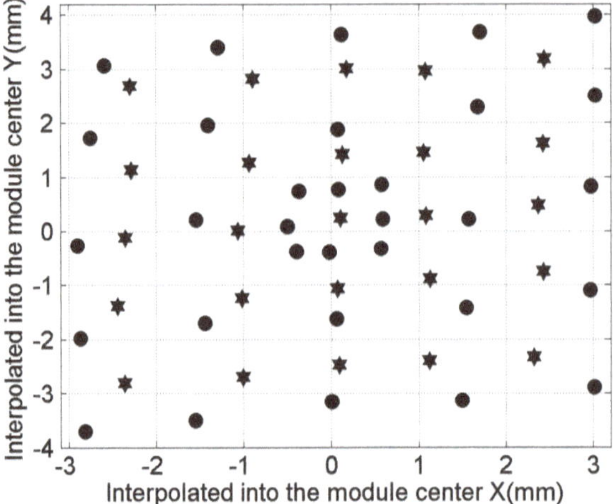

Fig. 5.10 Grid-like beam scan (calibration samples are in circles and validation samples are in hexagrams). The positions are interpolated into the center of the module

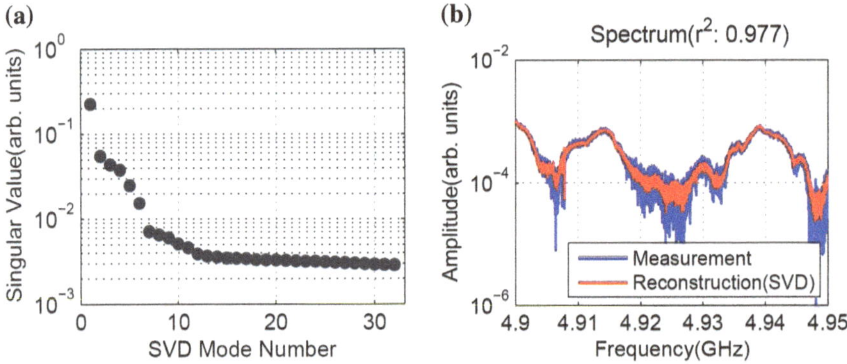

Fig. 5.11 **a** The singular value of each SVD mode. **b** Reconstructed spectrum (*red*) from the first 10 SVD modes compared with original spectrum (*blue*). Reprinted with permission from [1]. Copyright 2012, American Institute of Physics

As in the previous section, DLR, SVD and *k*-means clustering are applied. The singular values of all SVD modes decomposed from calibration samples are shown in Fig. 5.11a. The first 10 SVD modes have relatively large singular values, and therefore are used for the following analysis. The measured spectrum is compared with the reconstructed spectrum by using the first 10 SVD modes in Fig. 5.11b. These two spectra possess very good consistency, which indicates the representativeness of the first 10 SVD modes in describing the entire signal. The rms errors obtained with each method are presented in Table 5.1. The SVD prediction errors for validation samples are better than those of DLR and Clustering.

Table 5.1 Direct comparison of DLR, SVD and k-means clustering for the fixed sample split (Fig. 5.10). The signal is measured from HOM coupler C1H1

4.9–4.95 GHz	E_{RMS}^{cal} (x) (mm)	E_{RMS}^{cal} (y) (mm)	E_{RMS}^{val} (x) (mm)	E_{RMS}^{val} (y) (mm)
DLR	0.06	0.07	0.26	0.32
SVD	0.20	0.09	0.24	0.28
Clustering	0.20	0.17	0.36	0.34

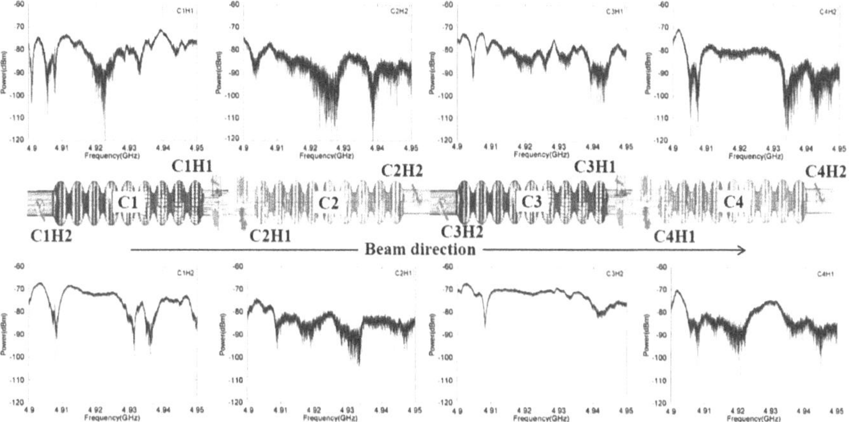

Fig. 5.12 Spectra measured from all eight HOM couplers of ACC39 module at one beam position (the *most top-right point* in Fig. 5.10)

In comparison to trapped cavity modes (Table 4.4), the coupled modes have larger prediction errors which is inconsistent with their stronger coupling to the beam. We believe this is coming from the fact that the signal to noise ratio is better for the trapped cavity modes due to the removal of the 10 dB external attenuator for each coupler during the data taking. This issue has been clarified in later measurements with the test electronics, as will be described in Chap. 6.

The same analysis scheme was extended to all other HOM couplers. The spectra of eight couplers ranging from 4.9 to 4.95 GHz are shown in Fig. 5.12 for one beam position. The obtained rms errors are shown in Fig. 5.13.

Based on the results shown in this chapter, three modal options promising for beam position diagnostics have been narrowed down: localized dipole beam-pipe modes at approximately 4 GHz, coupled cavity modes in the first two dipole bands within 4.5–5.5 GHz and trapped cavity modes in the fifth dipole band at approximately 9 GHz. A set of test electronics was subsequently built to study the resolution when predicting beam positions using these modal options. These will be described in Chap. 6.

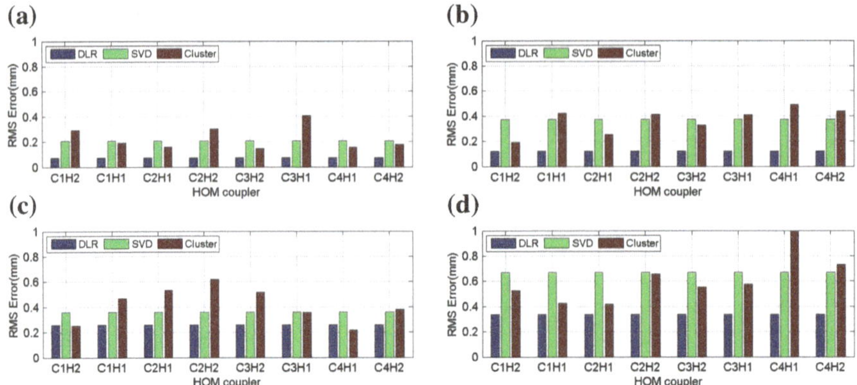

Fig. 5.13 The rms errors of all eight HOM couplers for spectra ranging from 4.9 to 4.95 GHz and split as Fig. 5.10. The applied method is DLR, SVD with the first 10 SVD modes and k-means clustering with 9 clusters. **a** x (Calibration). **b** y (Calibration). **c** x (Validation). **d** y (Validation)

References

1. P. Zhang, N. Baboi, R.M. Jones, I.R.R. Shinton, T. Flisgen, H.W. Glock, A study of beam position diagnostics using beam-excited dipole modes in third harmonic superconducting accelerating cavities at a free-electron laser. Rev. Sci. Instrum. **83**, 085117 (2012)
2. Fundamentals of Real-Time Spectrum Analysis, Tektronix Application Note, 2010
3. S. Molloy et al., High precision superconducting cavity diagnostics with higher order mode measurements. Phys. Rev. ST Accel. Beams **9**, 112801 (2006)
4. P. Zhang, N. Baboi and R.M. Jones, Higher order mode spectra and the dependence of localized dipole modes on the transverse beam position in third harmonic superconducting cavities at FLASH. DESY Report: DESY, pp. 12–109, 2012
5. N. Baboi et al., *Higher Order Modes for Beam Diagnostics in Third Harmonic 3.9 GHz Accelerating Modules*, in *Proceedings of SRF2011*, (Chicago, 2011), pp. 239–243
6. P. Zhang, N. Baboi and R.M. Jones, A Study of Beam Alignment Based on Coupling Modes in Third Harmonic Superconducting Cavities at FLASH, in DITANET International Conference: Accelerator Instrumentation and Beam Diagnostics (Seville, Spain, 2011). e-prints: arXiv:1111.3479
7. P. Zhang et al., *Beam-based HOM Study in Third Harmonic SC Cavities for Beam Alignment at FLASH*, in *Proceedings of DIPAC2011*, (Hamburg, Germany, 2011), pp. 77–79

Chapter 6
HOM-Based Beam Position Diagnostics

Following the comprehensive series of simulations and HOM measurements of the third harmonic 3.9 GHz cavities described in the previous chapters, we ascertained that there are three significant regions of interest useful for beam position diagnostics. The first region contains modes which are coupled from one cavity to the next within the ACC39 module. The next region is the fifth dipole band containing trapped cavity modes. The third region is the localized dipole beam-pipe modes. Each region has its own merits. The coupled cavity region, allows a superior position accuracy, whereas the trapped cavity modes or localized beam-pipe modes allow diagnostics on an individual cavity or beam pipe, but with a reduced accuracy compared to the coupled one. In order to be able to choose the frequency best suited for diagnostics, the three options were studied with the help of a test electronics, which overcomes the resolution limitation of the spectrum analyzer previously used [1, 2].

In this chapter, the principle of the test electronics is described in Chap. 6.1. Trapped cavity modes in the fifth dipole band provide local beam position inside the cavity, and are addressed in Chap. 6.2. Dimension reduction techniques along with a linear regression (described in Chap.4) have been used to extract the beam position from HOM signals. The dependence of the position resolution on various dipole modes and various time windows is studied. The resolution for a four-cavity-module-based beam position is presented in Chap. 6.3 using coupled cavity modes in the second dipole band. An analysis of contributors to the position resolutions is presented in Chap. 6.4.

6.1 The Principle of Custom-Built Test Electronics

A set of analog test electronics was designed at Fermilab in order to have the flexibility to study various modal options of interest as well as accommodating the large frequency bandwidths required, as explained in Chap. 5. Its simplified block diagram is shown in Fig. 6.1. One of four different analog bandpass filters (BPF) can be con-

P. Zhang, *Beam Diagnostics in Superconducting Accelerating Cavities*, Springer Theses, DOI: 10.1007/978-3-319-00759-5_6,

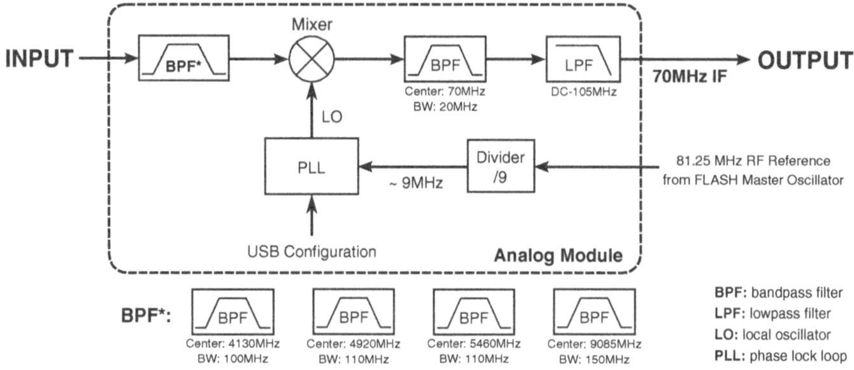

Fig. 6.1 Schematics of the analog electronics inside the box. One of the four BPFs was connected in front of the mixer during each measurement. Reproduced with permission from Fig. 4 of [1]

nected into the chain to study localized dipole beam-pipe modes, multi-cavity modes in the first and the second dipole band and trapped cavity modes in the fifth dipole band. For example, for the fourth BPF, the HOM signal has dominant components within 9010–9120 MHz[1]. Then the signal is mixed with a selectable local oscillator (LO) with a frequency set to 9136 MHz in this case. The 9066 MHz RF signal is therefore downmixed to an intermediate frequency (IF) of 70 MHz. The mixer is a single-side-band mixer with 26 dB image rejection on the higher frequency side. The LO frequency can be programmed in 9 MHz increments via a USB configuration and retains phase lock with the 9 MHz trigger signal by a Phase Lock Loop (PLL). This 9 MHz trigger is generated by dividing a 81.25 MHz reference signal from the FLASH master oscillator. The 70 MHz IF signal is further filtered with a 20 MHz analog BPF to study a specific band of modes. In order to ensure that possible remaining high-frequency components of the IF signal generated during the mixing step are well suppressed, a lowpass filter (LPF) is applied to preserve only the frequency below 105 MHz with dominant components from 60 to 80 MHz. A detailed circuit drawing of the analog electronics is shown in Fig. C.3.

In Fig. 6.1, the "INPUT" is connected to a RF multiplexer to allow for a fast switching among HOM couplers as shown in Fig. 6.2. The IF signal from the "OUTPUT" is then processed by a VME digitizer operating at 216 MS/s with 14 bit resolution[2] and 2 Volt peak-to-peak signal range along with a programmable FPGA for signal processing. The digitizer is triggered by a 10 Hz beam trigger, which is the RF-pulse repetition rate of FLASH (see Fig. 1.2). Both the selectable LO and the digitizer clock of 216 MHz are locked to the accelerator by using signals delivered from the

[1] This is the effective bandwidth obtained from measurements which is narrower than the BPF specifications.

[2] The output value from the VME digitizer has 16 bit with the last two digits both set to 0. It was for easy data processing with the FPGA. The effective resolution is 14 bit.

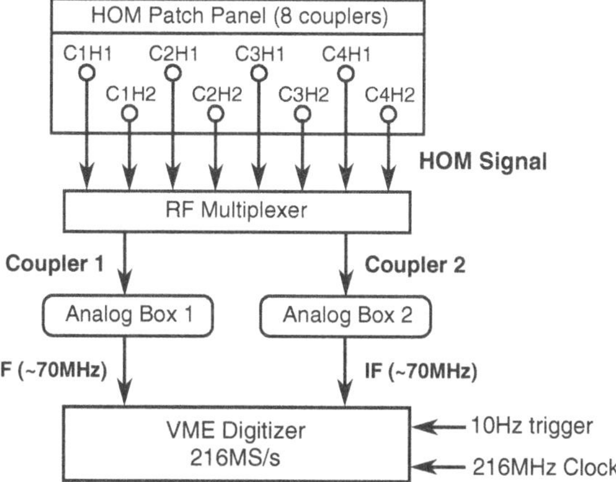

Fig. 6.2 Schematic device setup in the injector rack outside the FLASH tunnel

master oscillator as a reference. This locking allows correct phase information of the digitized signal.

Two sets of similar analog electronics have been built. All the devices are set up in the FLASH injector rack outside the accelerator tunnel (see Fig. 6.2). HOM signals of all eight couplers from the ACC39 patch panel are connected to the RF multiplexer. Signals from all four coupler-1's (H1) are switched to one analog electronics while coupler-2's to the other. Photos of the device setup, the analog electronics and the digitizer are shown in Fig. C.2. The digitized data is collected from the digitizer with the EPICS [3] software tool. For each measurement with the test electronics, the bunch charge, currents of the steering magnets and BPM readouts are recorded synchronously from the FLASH control system DOOCS [4], similar to the measurements with the RSA presented in Chap. 5. The data processing for position diagnostics is performed offline using MATLAB [5].

Example waveforms of the electronics' output when connecting coupler C3H2 to its input are shown in Fig. 6.3 for trapped and coupled modes. Each of them is excited by a single electron bunch. HOMs are excited at approximately 100th sampling point, and then decay in a relatively short time (approximately 1.4 μs for Fig. 6.3a and 0.7 μs for Fig. 6.3b).

6.2 Position Diagnostics with Trapped Cavity Modes

First, data measured from HOM coupler C3H2 are presented. The fourth bandpass filter in Fig. 6.1 was connected to obtain the fifth dipole band at approximately 9 GHz. The LO has been set to down-convert 9066 MHz to 70 MHz. An example digitizer output is shown in Fig. 6.3b. Similar to the analysis of the RSA data in Chap. 4, three analysis methods (DLR, SVD and k-means clustering) are applied on the waveforms to extract beam position information. A cross-validation technique has also been

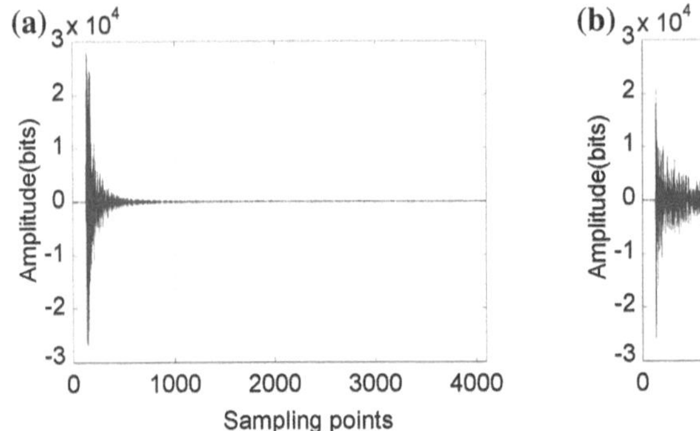

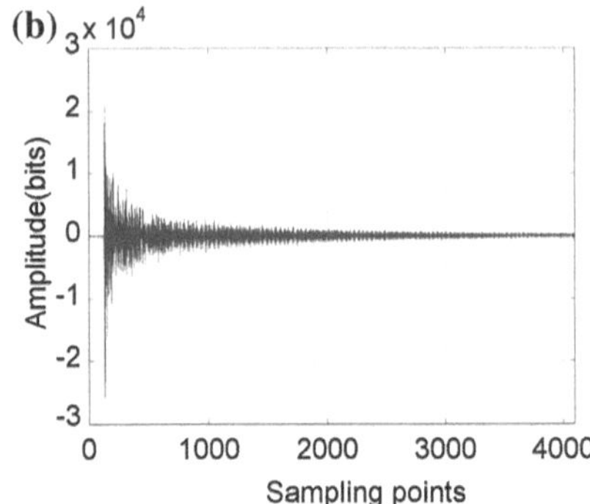

Fig. 6.3 (**a**) Center: 5437 MHz. (**b**) 1wfm-raw-5437 MHz-D208. Signals of the digitizer output for two different bands from coupler C3H2. The length of the waveform is 4096 sampling points corresponding to approximately 19 μs. Reproduced with permission from Fig. 5 of [1]

used to evaluate the position resolution. The dependence of position resolution on various dipole modes and various time windows is studied. Finally, the distribution of a sum of modal energies within this band and position resolutions by using HOMs extracted from each coupler are depicted.

6.2.1 Data Preparation

The frequency components of the waveform (Fig. 6.3b) can be obtained by performing a FFT (Fast Fourier Transform). The magnitude of the FFT signal is shown in Fig. 6.4a. From previous studies described in Chap. 3 (Fig. 3.10c), we know that modes below the double peaks at approximately 55 MHz are propagating through cavities, therefore, an ideal filter has been applied mathematically on the waveform to cut away the frequency components below 58 MHz and above 96 MHz. Only the region within the two dash lines in Fig. 6.4a has been preserved. Since the signal decays very fast (see Fig. 6.4b), a time window has been applied on the filtered waveform to reduce the signal length. The final waveform has a length of 700 sampling points (corresponding to approximately 3.2 μs) and has been normalized to the bunch charge as shown in Fig. 6.4c. This will be used in the following study.

The measurement setup is the same as described in Chap. 5 (see Fig. 5.1). The beam was moved in a 2D-grid manner by using steering magnets located upstream of the ACC1 module as shown. We have switched off the non-linear elements from the RF gun up to BPM-B so as to create a straight-line beam trajectory. The RF of ACC1 was left on due to difficulties to transport the beam at 5 MeV. As explained in Chap. 5, this however induces significant technical difficulties to move the beam in a random angle. BPM-A and BPM-B were used to record the transverse beam position before and after the bunch entering ACC39. The FLASH machine was operated with

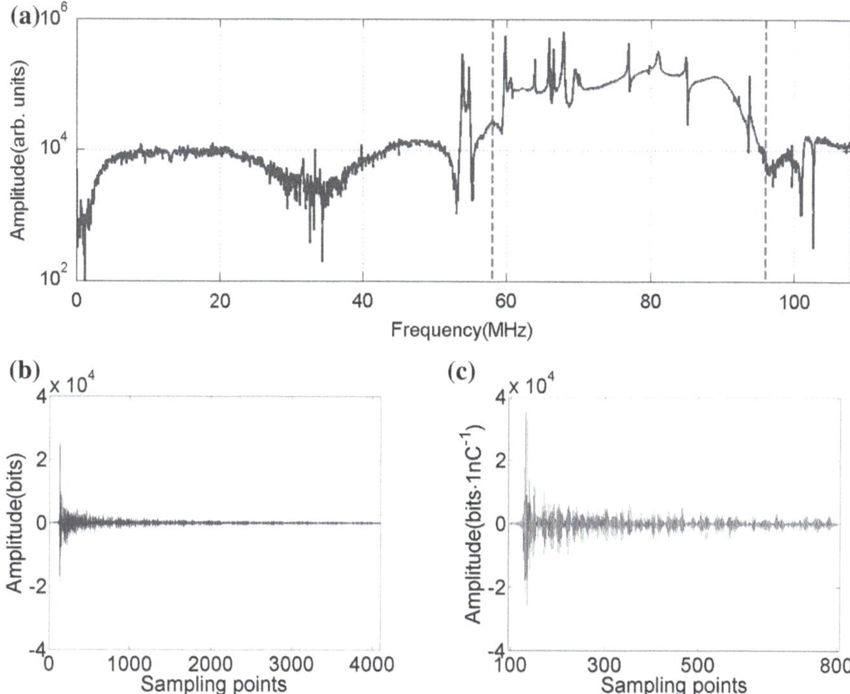

Fig. 6.4 (**a**) Frequency components of the digitized signal in Fig. 6.3b. (**b**) Time-domain waveform after applying the ideal filter. (**c**) Filtered waveform with a time window and normalized to the bunch charge. Reproduced with permission from Fig. 6 of [1]

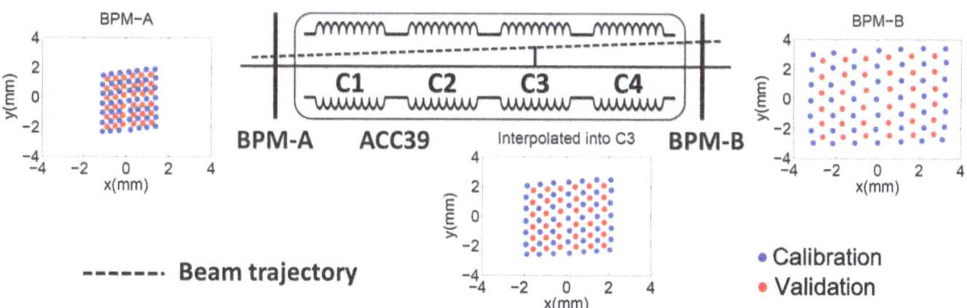

Fig. 6.5 Position interpolated into the center of C3 by readouts of BMP-A and BPM-B. Reproduced with permission from Fig. 7 of [1]

only one bunch in each macropulse. The bunch charge was approximately 0.5 nC recorded from a toroid situated right after BPM-B. A position interpolation is applied to obtain the transverse beam position in each cavity as shown in Fig. 6.5 for C3 as an example. These interpolated positions are used in the following study. We first took 49 calibration samples (blue dots in Fig. 6.5) and then 36 validation samples (red dots in Fig. 6.5).

6.2.2 Extraction of Beam Position from HOM Waveforms

The analysis procedure we used to determine the correlation between HOM signals and beam positions is similar to the previous studies with the RSA. The difference lies in the HOM signals used: waveforms are used in this chapter instead of spectra used in Chap. 5. Three different methods are applied: DLR, SVD and k-means clustering. The details of using these methods are described in Chap. 4. SVD and k-means clustering are used to reduce the dimension of the linear system and suppress noises present in the HOM waveforms.

For the SVD method, the singular value of each SVD mode decomposed from calibration samples (blue dots in Fig. 6.5) is shown in Fig. 6.6a. The first several SVD modes have relatively large singular values, therefore their SVD mode amplitudes are used to regress on the interpolated beam positions instead of the original HOM waveforms. A calibration of the HOM signals is generated in this manner. The predictions of beam positions are made by applying the calibration on validation samples. These predictions are then compared with the measured position interpolated into the cavity. The rms of the position errors (rms error) for validation samples is defined as the position resolution (Appendix A.4). The contribution of using the first p SVD modes to determine the beam position is shown in Fig. 6.6b for x and Fig 6.6c for y. The first 12 SVD modes provide a surprisingly small rms error in both x and y for our purposes.

For the k-means clustering methods, the contributions of using k clusters to determine the beam position are shown in Fig. 6.7. 18 clusters suggest an optimal rms error in both x and y.

The predictions of beam positions obtained by DLR, SVD with the first 12 SVD modes and k-means clustering with 18 clusters are compared with measured beam positions in Fig. 6.8. The coefficient of determination r^2 (Appendix A.3) is also given. The prediction errors are shown in Fig. 6.9. The position resolutions are calculated and listed in Table 6.1. The rms errors are comparable for DLR and SVD of both calibration and validation samples. Both methods suggest an approximately 50 μm position resolution. The k-means clustering suggests a worse resolution of 60–70 μm.

6.2.3 Cross-Validation

Until now, our analysis is based on a specific sample split (Fig. 6.5). To remove sample dependence, a technique namely *cross-validation* is used [6]. As in the discussed example the total sample size is only 85 (49 calibration positions plus 36 validation positions), one can apply the *leave-one-out cross-validation* (LOOCV) technique [6]. LOOCV uses only one from the total samples for the validation and the remaining as calibrations. This is repeated for every sample (85 times in the considered case). Fig. 6.10 shows the measurement and the prediction of each validation sample. The applied methods are DLR (red), SVD (green) with the first 12 SVD modes and k-means clustering (magenta) with 18 clusters. The rms errors are then calculated

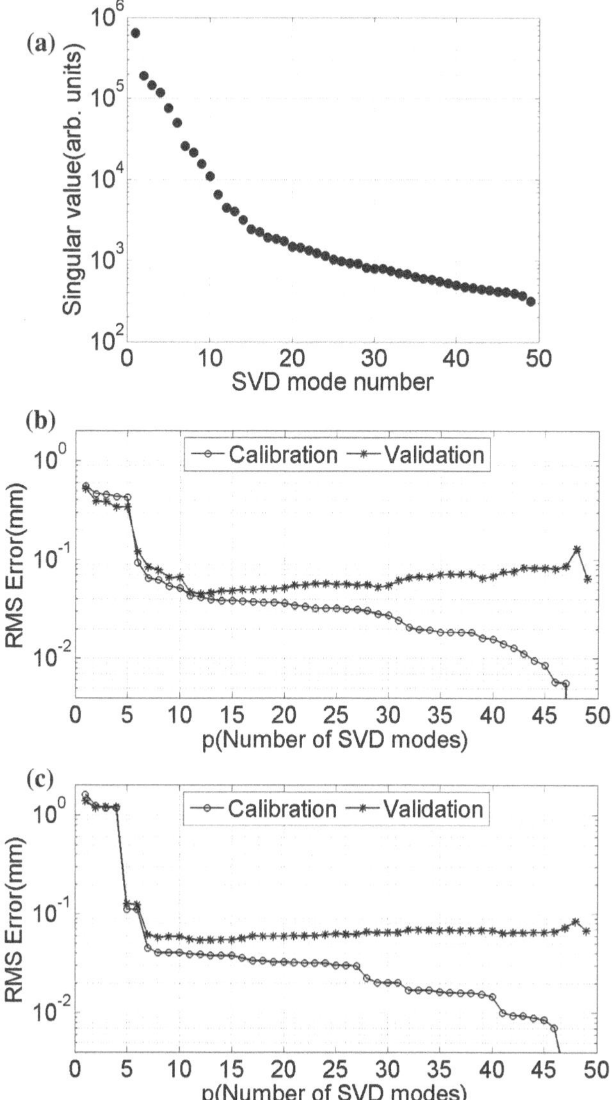

Fig. 6.6 (**a**) Singular value of each SVD mode from calibration samples. (**b**) (**c**) Contributions of the first p SVD modes to determine the transverse beam position x and y measured by the rms error. Reproduced with permission from Fig. 8 of [1]

on the prediction errors for the validation sample from all 85 different sample splits and these are shown in Fig. 6.11. The sample-independent rms errors are listed in Table 6.2. Since the calibration samples for each of the 85 splits are similar, the sample-independent rms error for the validations is a good estimation of the position

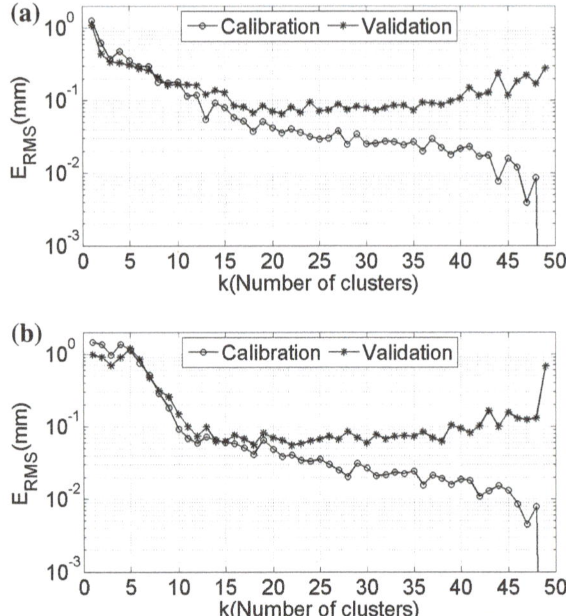

Fig. 6.7 (**a**) rms error vs. k (x). (**b**) rms error vs. k (y). Contributions of k clusters to determine the transverse beam position x and y measured by the rms error

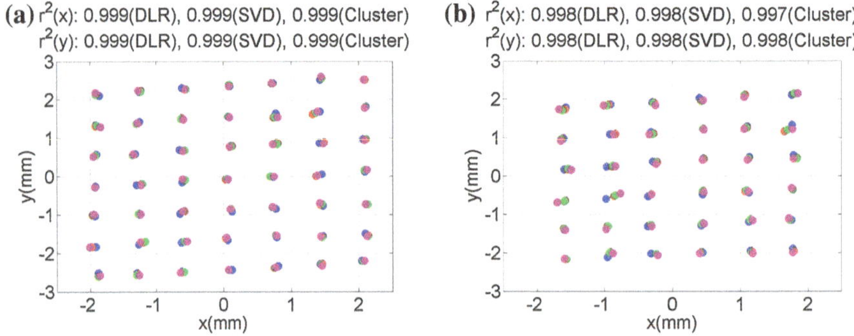

Fig. 6.8 (**a**) Calibration. (**b**) Validation. Measurements (*blue*) and predictions of the transverse beam position from calibration and validation samples. The DLR is in *red*. The SVD is in *green* with the first 12 SVD modes. The k-means clustering is in *magenta* with 18 clusters

resolution. The results are similar compared with the fixed sample split case shown in Table 6.1. Although the results are comparable for all three methods, SVD and k-means clustering are more efficient than DLR in computing calibration coefficients, since the number of unknown variables is much smaller, as shown in Table 6.2. However, once the calibration matrix M is obtained, the time required for position prediction is generally equally fast.

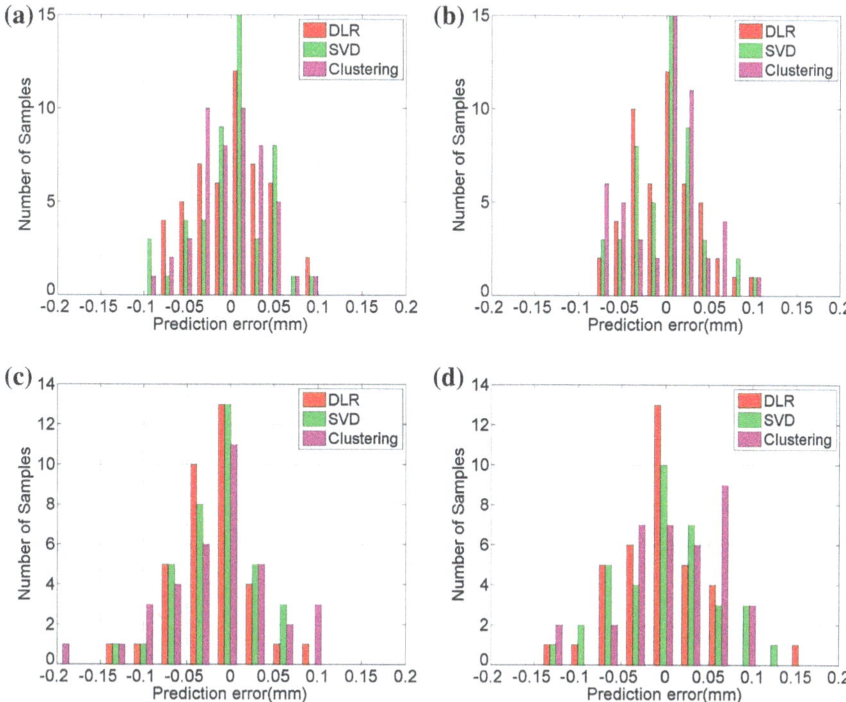

Fig. 6.9 (**a**) x (Calibration). (**b**) y (Calibration). (**c**) x (Validation). (**d**) y (Validation). Prediction errors of the transverse beam position from calibration and validation samples. The DLR is in *red*. The SVD is in *green* with the first 12 SVD modes. The k-means clustering is in *magenta* with 18 clusters

Table 6.1 Direct comparison of DLR, SVD and k-means clustering for the fixed sample split (Fig. 6.5). The signal was measured from HOM coupler C3H2

Fixed Split	E_{rms}^{cal} (x) (μm)	E_{rms}^{cal} (y) (μm)	E_{rms}^{val} (x) (μm)	E_{rms}^{val} (y) (μm)
DLR	41	40	47	56
SVD	42	39	45	54
Clustering	37	41	67	56

6.2.4 The Search for Suitable Trapped Dipole Modes

Detailed simulations (Appendix B) have shown that modes in the fifth dipole band couple weakly to the beam. This will affect the resolution of beam positions in the cavity measured with these dipole modes. In order to improve the resolution, we use a band of modes for position determination as shown in the previous section. This is different from the previous study in Ref. [7], which uses two polarizations of one dipole mode. In Fig. 6.12a, resolutions using a band of modes are shown as

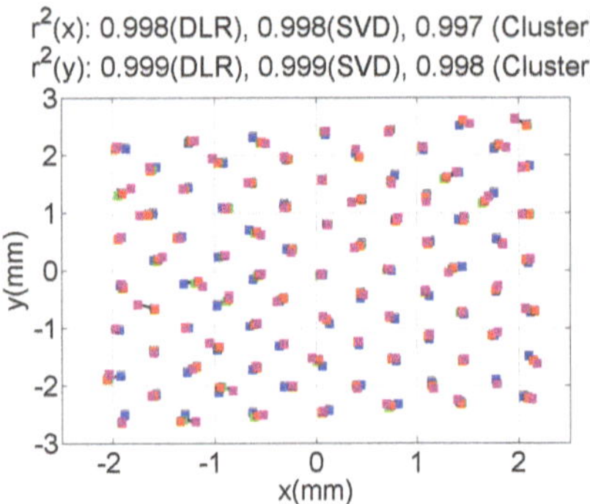

Fig. 6.10 Measurement (*blue*) and prediction of the transverse beam position from each of 85 validation samples using LOOCV. The DLR is in *red*. The SVD is in green with the first 12 SVD modes. The k-means clustering is in *magenta* with 18 clusters. Points connected with *black* lines belong to the same beam position

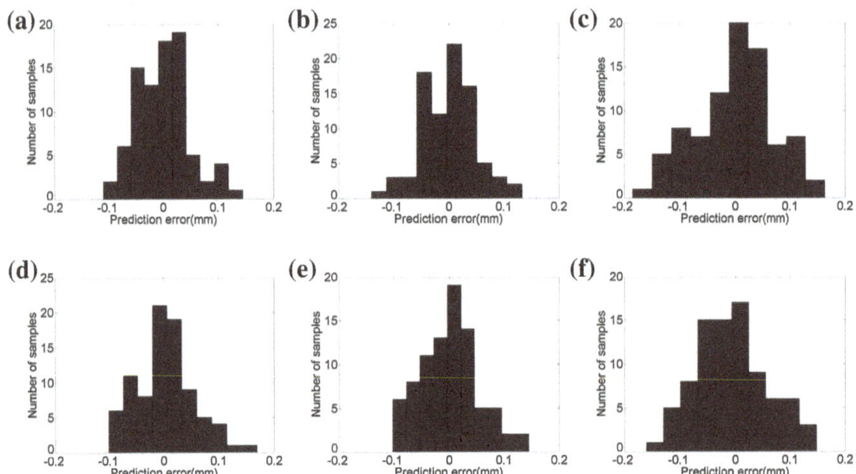

Fig. 6.11 (**a**) x (DLR). (**b**) x (SVD). (**c**) x (Clustering). (**d**) y (DLR). (**e**) y (SVD)(f)y (Clustering). Prediction errors of the transverse beam position from each of 85 validation samples using LOOCV

the long magenta lines on the bottom. These are compared with resolutions using different segments of the band. The color of the segment in Fig. 6.12b matches the color of the resolution lines in Fig. 6.12a. An ideal filter is applied mathematically on the time-domain digitizer output (Fig. 6.3b) using the bandwidth of each segment in Fig. 6.12b. A time window is then applied on the filtered waveform to truncate the signal properly. Fig. 6.12a reveals a remarkable improvement in resolution achieved through combining modes in the trapped band.

Due to the 3.9 GHz cavity being smaller than the 1.3 GHz cavity in size, eigenmodes of ACC39 are particularly sensitive to small geometrical errors. Predicting

Table 6.2 Direct comparison of DLR, SVD and k-means clustering using cross-validation

	E_{rms} (x) μm	E_{rms} (y) μm	Number of unknowns
DLR	50	52	702
SVD	50	52	13
Clustering	70	65	19

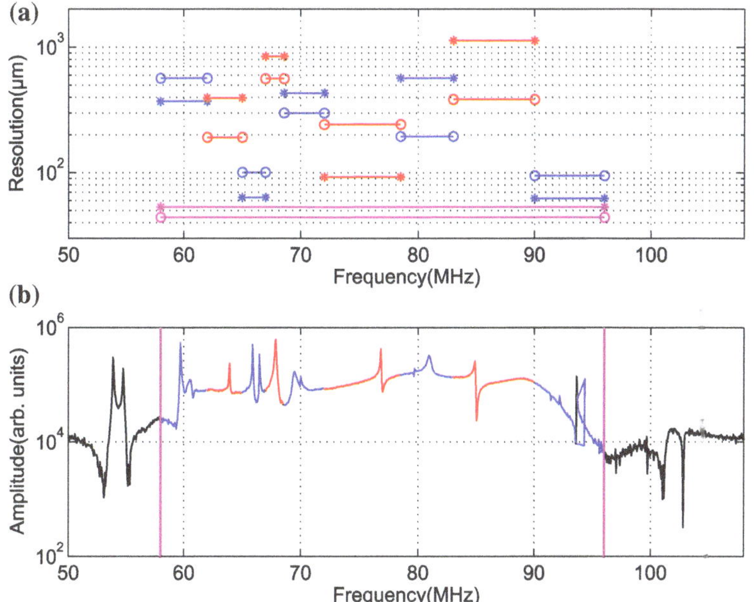

Fig. 6.12 Resolutions of each segment of spectrum. Circles denote x and asterisks denote y. Reproduced with permission from Fig. 10 of [1]

the exact frequency locations of these modes is impractical. For this reason we utilize the signal of a complete band of frequencies to accommodate the frequency shifts of dipole modes extracted from different cavities.

6.2.5 The Search for a Suitable Time Window

HOMs in the fifth dipole band decay rapidly as shown in Fig. 6.3b. In this section, we study the resolution dependence on the time window applied on waveforms. The method we used is SVD. Starting from the 100th sampling point, the length of the time window increases in the forward direction by a step of 50 sampling points. This is illustrated in Fig. 6.13a. The waveforms with each time window are then treated with SVD and various number of SVD modes are subsequently used for

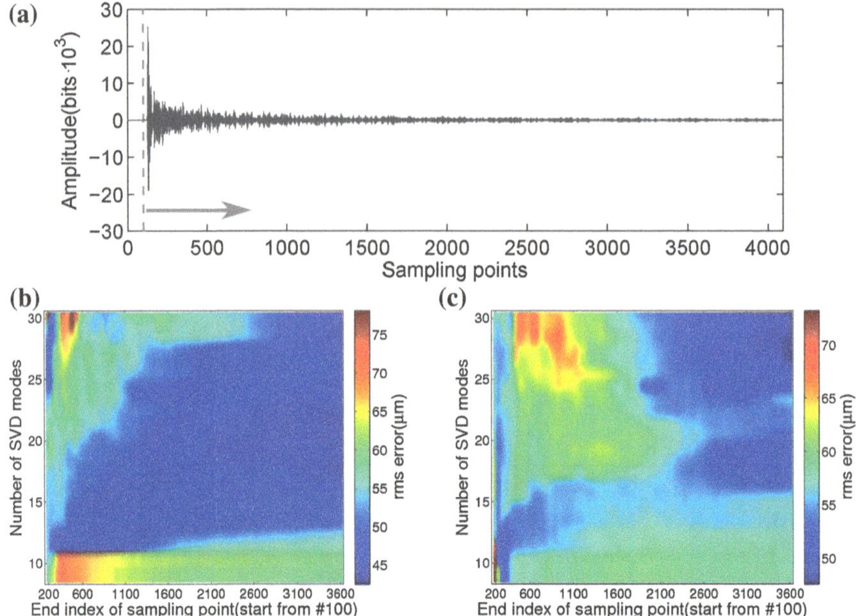

Fig. 6.13 (**a**) Forward scan. (**b**) x. (**c**) y Scan forward starting from the 100th sampling point. Reproduced with permission from Fig. 11 of [1]

position determinations. The resolutions are shown in Fig. 6.13b and c for different time window and different number of SVD modes. In Fig. 6.13b, a minimum length of time window starting from the 100th and ending at the 300th sampling point with 11 SVD modes gives a reasonably good resolution of 47 μm for x. The same time window with 9 SVD modes suggest a resolution of 49 μm for y as shown in Fig. 6.13c.

Having found the smallest end index, the 300th, of the time window from the forward scan, we conduct a backward scan by moving the start index of the sampling point in the backward direction with a step of 10 sampling points. This is shown in Fig. 6.14a. A shorter time window from the 140th to the 300th with 11 SVD modes give a same resolution of 47 μm for x as shown in Fig. 6.14b. A resolution of 51 μm for y is expected by using the same time window with 9 SVD modes as shown in Fig. 6.14c.

The findings from the forward and backward scans are summarized in Table 6.3. The length of the time window is also calculated in terms of time. As a comparison, the resolutions with a time window from the previous section are also listed. In an extreme case, the full length of the waveforms is also treated in the same way and resolutions are calculated. From the table, we conclude that a small length of waveform contains enough information to determine beam position with reasonably good resolutions. The first 40 sampling points (from the 100 to 140th, corresponding

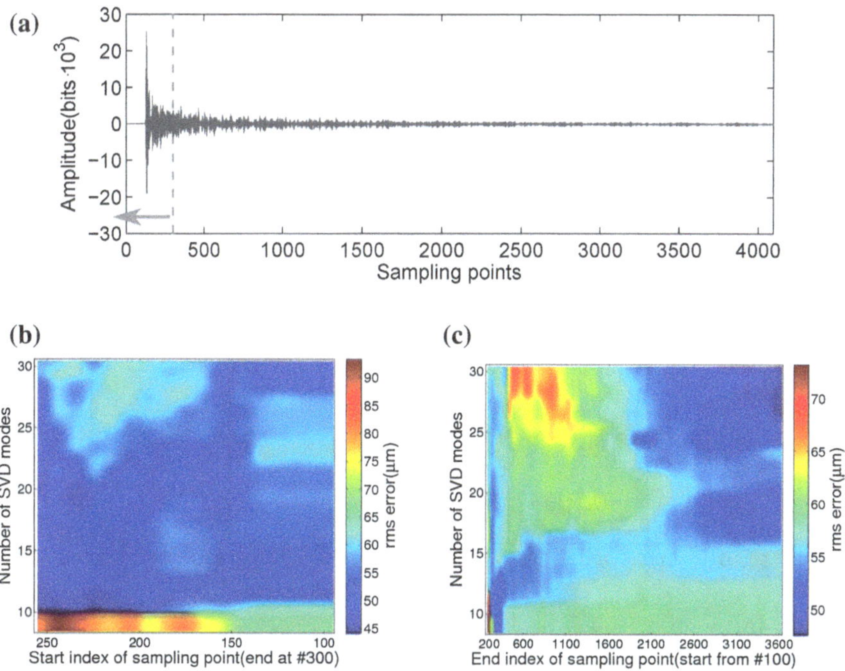

Fig. 6.14 (**a**) Backward scan. (**b**) x. (**c**) y. Scan backward ending at the 300th sampling point. Reproduced with permission from Fig. 12 of [1]

to approximately 0.19 μs) which have large amplitudes appear to be insensitive to beam movement. This might be due to a beam pulse effect.

As shown in Table 6.3, the backward scan suggests a time window from 140 to 300th with a length of 0.7 μs. This length is shorter than the bunch separation of 1 μs at FLASH when operated in a multi-bunch mode (see Fig. 1.2). Figure 6.15 shows the waveform with different time windows. The black lines represent the multiple bunches separated by 1 μs, while the red dash lines define the shortest time window found with reasonably good position resolutions. This suggests non-degenerate position resolutions of the leading bunch in a bunch train. Although HOMs decay fast, there is still a remaining signal from the current bunch overlapping with HOMs excited by trailing bunches. This will dilute the beam position resolution of bunches following the leading bunch in a train at FLASH. However, this does not diminish the usefulness of this diagnostics, as degenerate beam positions of trailing bunches are also desirable.

Table 6.3 Resolution of various segments of waveform with different number of sampling points. Reproduced with permission from Table 1 of [1]

	Sampling points			Resolution		SVD modes	
	start #	end #	t(μs)	x(μm)	y(μm)	x	y
Forward scan	100	300	0.9	47	49	1st–11th	1st–9th
Backward scan	140	300	0.7	47	51	1st–11th	1st–9th
Full segment	100	800	3.2	45	54	1st–12th	1st–12th
Full waveform	1	4096	18.9	43	45	1st–17th	1st–25th

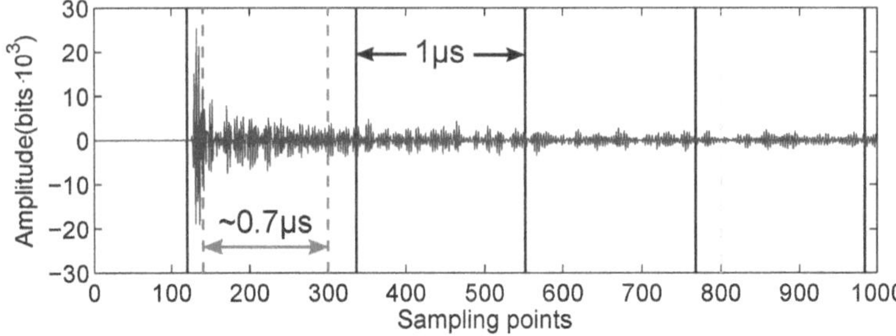

Fig. 6.15 Waveform with various time windows. Reproduced with permission from Fig. 13 of [1]

6.2.6 Results for all HOM Couplers

The integrated power over the frequency region 58–96 MHz (see Fig. 6.4a) is calculated for each beam position. Fig. 6.16 shows the integrated power distribution for each HOM coupler. Neither the power distribution nor the power minimum is similar among couplers. This might be a proof that modes are localized inside each cavity presented in the frequency region for power integration. It might also be an implication of different electrical axes for individual cavities. This difference can be attributed to asymmetric structure due to couplers and fabrication tolerances of cavities. In addition, dissimilar couplers with different coupling to HOMs can also contribute.

The resolution of predicting the transverse beam position inside each cavity by using the HOM signal extracted from each of the eight HOM couplers is obtained by SVD with the first 12 SVD modes. This is shown in Fig. 6.17. A resolution of 40–60 μm for x and 50–100 μm for y are suggested.

6.3 Position Diagnostics Based on a Band of Coupled Modes

Previous studies (Chap. 3) have shown that modes in the first two dipole bands propagate amongst all cavities within the ACC39 module. This enables the beam position

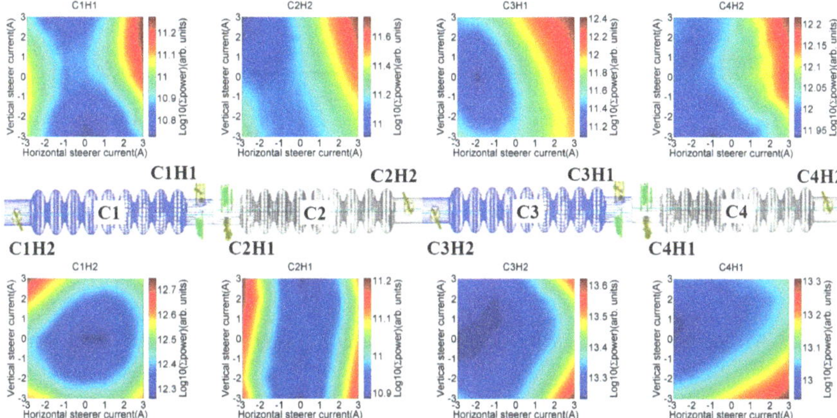

Fig. 6.16 Integrated power (log scale) as a function of steering magnets' current

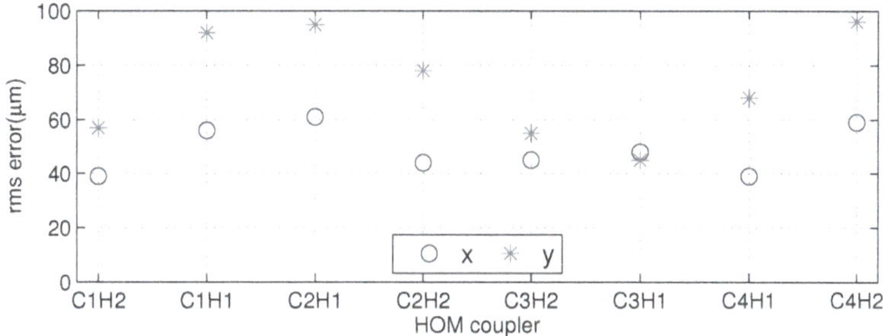

Fig. 6.17 Position resolution for each coupler obtained by SVD with the first 12 SVD modes

to be determined in a four-cavity module sense rather than for each individual cavity. Therefore, we interpolate the two BPM readouts into the center of the module (see Fig. 6.18) and study the beam position resolution at this location. Simulations predict some strong coupling modes (large R/Q values) in the first two dipole bands (Appendix B). HOM BPM using these modes will potentially have high resolutions. At first, modes in the second dipole band measured from coupler C3H2 are shown. The third BPF in Fig. 6.1 was connected. The LO has been set to downconvert 5437 to 70 MHz. An example digitizer output is shown in Fig. 6.3a. The frequency components obtained by FFT are shown in Fig. 6.19a. A time window has been applied to cut away the saturated part of the waveform at the beginning of the signal (cut at the 170th sampling point), as well as the trailing part due to a fast signal decay (end at the 500th sampling point). The resulting waveform is then normalized to the beam charge and is shown in Fig. 6.19b. These waveform slices are used to determine beam positions at the center of the module.

Predictions of beam positions are obtained by DLR, SVD with the first 11 SVD modes and k-means clustering with 13 clusters. Prediction errors are shown in

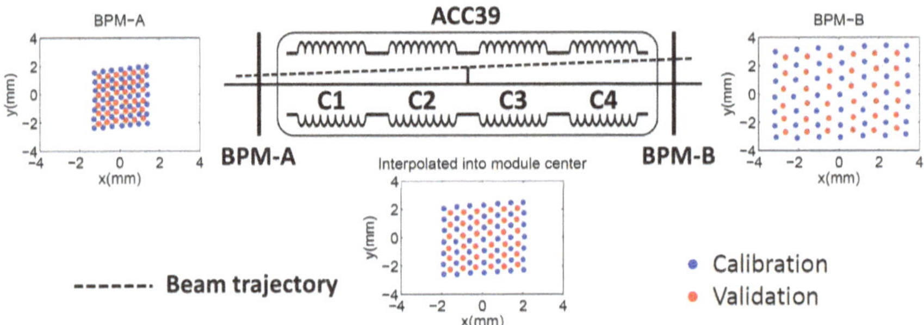

Fig. 6.18 Postion interpolated into the center of the four-cavity module by the readouts of BMP-A and BPM-B

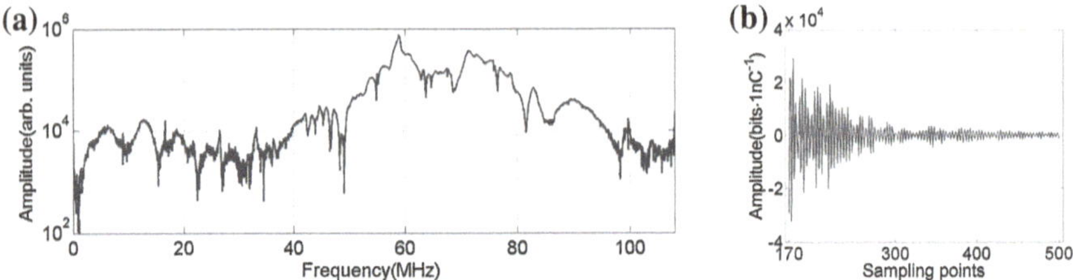

Fig. 6.19 (a) Frequency components of the digitized signal. (b) Waveform with a time window and normalized to the beam charge. Reproduced with permission from Fig. 14 of [1]

Table 6.4 Direct comparison of DLR, SVD and k-means clustering for the fixed sample split (Fig. 6.18). The signal was measured from HOM coupler C3H2

Fixed Split	E^{cal}_{rms} (x) μm	E^{cal}_{rms} (y) μm	E^{val}_{rms} (x) μm	E^{val}_{rms} (y) μm
DLR	19	17	20	25
SVD	18	13	20	22
Clustering	19	14	23	21

Fig. 6.20. The position resolutions are calculated and listed in Table 6.4. The rms errors are comparable for all three methods of both calibration and validation samples. All three methods suggest an approximately 20 μm position resolution.

To study the electromagnetic energy distribution of these coupled modes, the HOM power is integrated over a frequency range from 1 to 107 MHz (see Fig. 6.19a) for each beam position. The power distribution of these coupled modes extracted from each HOM coupler is shown in Fig. 6.21. A similar position with minimum power can be found among all eight couplers although the exact distribution for each coupler is dissimilar. This might be due to different orientations of the couplers and different types, therefore individual coupling to the HOMs. The longitudinal electromagnetic field distribution of these coupled modes can also contribute. Around this common position with minimum HOM power, a smaller scan was conducted as shown in

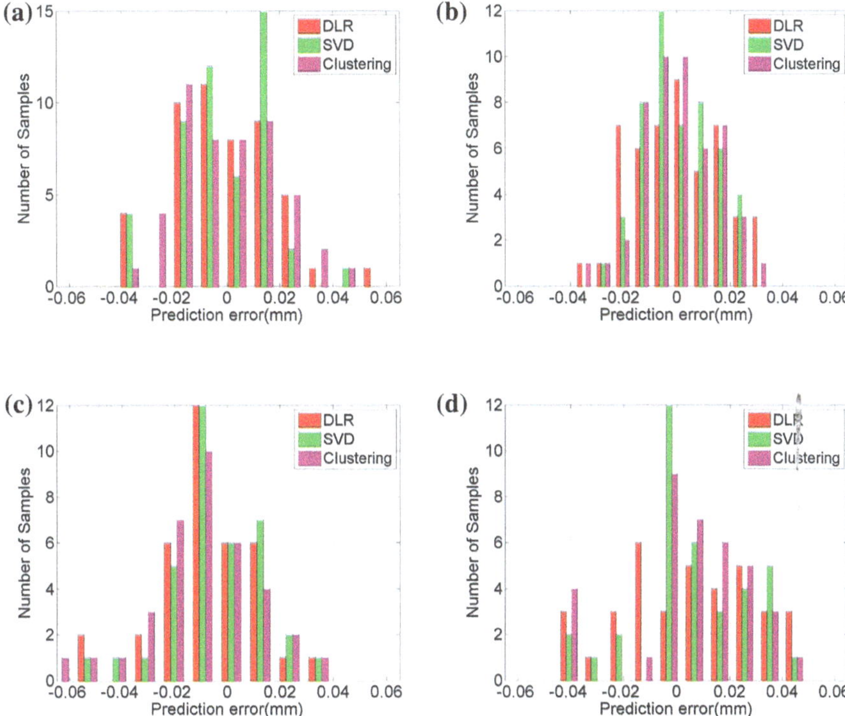

Fig. 6.20 (**a**) x (Calibration). (**b**) y (Calibration) (**c**) x (Validation). (**d**) y (Validation). Prediction errors of the transverse beam position from calibration and validation samples. The DLR is in red. The SVD is in green with the first 11 SVD modes. The k-means clustering is in magenta with 13 clusters

Fig. 6.22a. The resolution of predicting the transverse beam position at the center of the four-cavity module is obtained by using the HOM signal extracted from each of the eight HOM couplers. This is shown in Fig. 6.22b. Our study predicts a resolution of 10–25 μm for x and 20–30 μm for y.

Both localized and coupled cavity modes have been studied in predicting beam position with the test electronics. The present experiments suggest a resolution of 50 μm accuracy in predicting local beam position in the cavity and a global resolution of 20 μm over the complete module. Based on these results, the design of an online HOMBPM system for the 3.9 GHz module has been finalized. Two couplers will be equipped with electronics centered at 5460 MHz to provide high resolution positions in the module, while six couplers with electronics centered at 9060 MHz to deliver local position inside each cavity. In the next section, an estimation of contributors to the achieved position resolution is presented.

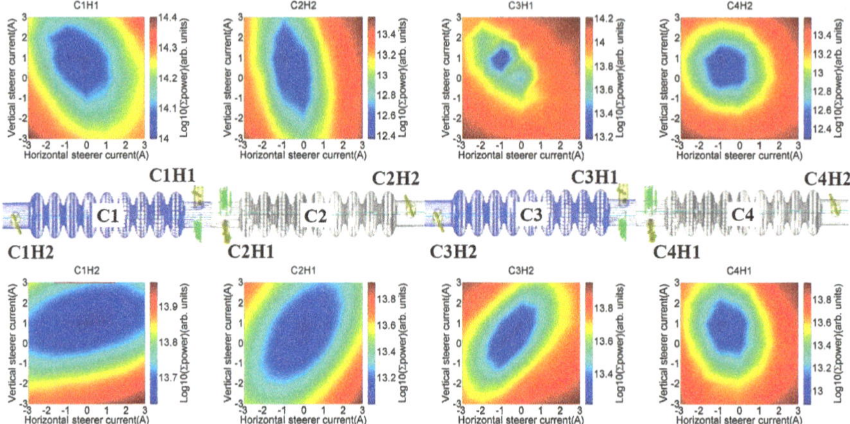

Fig. 6.21 Integrated power (log scale) as a function of steering magnets' current. Reproduced with permission from Fig. 16 of [1]

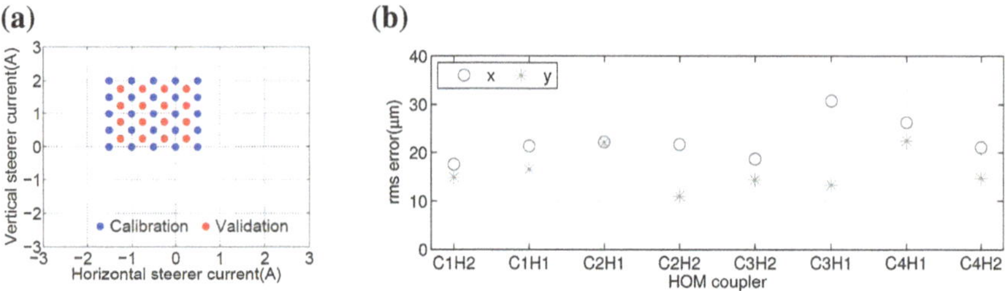

Fig. 6.22 Small scan **a** and the corresponding position resolution for each coupler **b**. Reproduced with permission from Fig. 17 of [1]

6.4 Fundamental Limitations to Position Resolution

The position resolution presented in previous sections receives contributions from many factors, which are estimated in this section. Fundamentally, the measurement resolution is limited by thermal noise, which is of course unavoidable at a non-zero temperature. The smallest measurable thermal energy, U_{th}, is [8]

$$U_{th} = \frac{1}{2} k_B T, \tag{6.1}$$

where T is the temperature and k_B is the Boltzmann constant (8.6×10^{-5} eV $\cdot$ K^{-1}). Assuming a room temperature of 300 K, $U_{th} = 0.0129$ eV.

A beam with charge q traversing a cavity loses an energy U in a particular mode as [9]

$$U = \frac{\omega}{2} \cdot \left(\frac{R}{Q}\right) \cdot q^2, \tag{6.2}$$

where R/Q defined in Chap. 1 has the unit of Ω/cm^2 for dipole modes. A fraction, β, of this energy radiates to the HOM coupler, where we use $\beta \sim 0.5$, following Ref. [7].

For the third harmonic 3.9 GHz cavity, the modes with the strongest coupling in the second dipole band has a simulated frequency of 5.4427 GHz and an R/Q value of 20.877 Ω/cm^2 (see Appendix B). The energy deposited in the cavity from a 1 nC beam is

$$U = \beta \cdot \frac{2\pi \cdot 5.4427\,\text{GHz}}{2} \cdot \frac{20.877\Omega/\text{cm}^2 \cdot (1\text{nC})^2}{1.602 \times 10^{-19}\text{C}} = 1.1 \times 10^{12}\ \text{eV}/\text{cm}^2, \tag{6.3}$$

where 1.602×10^{-19}C is the charge of $1e$. This corresponds to a transverse beam offset

$$r = \sqrt{\frac{U_{th}}{U}} = 1\,\text{nm}. \tag{6.4}$$

HOM signals are brought from the coupler inside the FLASH tunnel to the HOM patch panel outside the tunnel by long cables. The measured cable loss is approximately 20 dB, the electronics attenuation is 26 dB [2], and this degrades the resolution by a factor of f_{cable} as

$$r^* = f_{\text{cable}} \cdot r = 10^{\frac{20dB+26dB}{20}} \cdot r = 199.5 \times 1\,\text{nm} = 200\,\text{nm}. \tag{6.5}$$

The theoretical minimum resolution is therefore 200 nm, as compared to the observed resolution of 10–30 μm (see Fig. 6.22). This discrepancy may be due to four main issues.

First, HOM signals are all normalized to the bunch charge before the position prediction. The bunch charge was recorded from the nearby toroid, which has a measured resolution of 3 pC (rms). The measured bunch charge was approximately 0.5 nC throughout our studies, therefore it corresponds to a relative resolution of ~0.6 %. At 1 mm beam offset, this contributes ~6 μm to the position resolution.

Second, readouts of two BPMs are used to interpolate the beam position into the cavity/module and then to calibrate the HOM signals. These two BPMs have a measured resolution of ~10 μm (rms) [10], contributing to the resolution of the HOM-based system.

Third, the phase noise of the LO used to down-mix the HOM signal can also contribute to the resolution. This need to be studied in future measurements.

Fourth, the resolution we have obtained is in a global sense, since the rms is calculated on the prediction errors for all different beam offsets. Only one prediction is made for each individual beam position. This is a characterization of the prediction power for the entire beam scan, which is not usually used in beam diagnostics where resolution is quoted for a specific beam position [10]. In this case, the position varies mainly due to beam jitter from bunch to bunch. Initial studies on measuring beam jitter for specific beam offsets suggest improved resolutions by at least a factor of 2 [11].

Therefore, it can be seen that the measured position resolution is dominated by the BPM resolution used for position interpolation. For the measurement of local position inside each cavity, BPMs can be used to calibrate HOM signals in C1 and C3, the position measured in these two cavities can then be used to make a prediction about the position in C2. The details of this technique can be found in Ref. [10]. In this case, the limitations of BPMs are eliminated from the position prediction in C2, and might consequently improve the resolution.

References

1. P. Zhang, N. Baboi, R.M. Jones, N. Eddy, Resolution study of higher-order-mode-based beam position diagnostics using custom-built electronics in strongly coupled 3.9 GHz multi-cavity accelerating module. J. Instrum. **7**, P11016 (2012)
2. P. Zhang, N. Baboi N. Eddy, B. Fellenz, R.M. Jones, B. Lorbeer, T. Wamsat, M. Wendt, A study of beam position diagnostics with beam-excited dipole higher order modes using a downconverter test electronics in third harmonic 3.9 GHz superconducting accelerating cavities at FLASH, DESY Report: DESY 12–143 (2012)
3. EPICS. http://www.aps.anl.gov/epics/
4. DOOCS. http://tesla.desy.de/doocs/
5. MATLAB. Ver. R2011b, The MathWorks Inc., USA, http://www.mathworks.com/
6. R. Kohavi, A study of cross-validation and bootstrap for accuracy estimation and model selection, in *Proceedings of the 14th International Joint Conference on Artificial Intelligence*, vol. 2 (Montreal, Canada, 1995), pp. 1137–1143
7. S. Molloy et al., High precision superconducting cavity diagnostics with higher order mode measurements. Phys. Rev. ST Accel. Beams **9**, 112801 (2006)
8. W.B. Davenport, W.L. Root, *An introduction to the theory of random signals and noise*, (IEEE Press, 1987), ch. 9–4, p. 185
9. A. Chao, M. Tigner, *Handbook of Accelerator Physics and Engineering*, ch. 3.2.7, p. 213. Singapore: World Scientific, 1st edn., 1999
10. N. Baboi et al., Resolution studies at beam position monitors at the FLASH facility at DESY, in *Proceedings of BIW06* (Batavia, IL, USA, 2006)
11. T. Wamsat, N. Baboi, B. Lorbeer and P. Zhang, Performance of a downconverter test-electronics with TCA-based digitizers for beam position monitoring in 3.9GHz accelerating cavities, in *Proceedings of IBIC12*, (Tsukuba, Japan, 2012)

Chapter 7
Conclusions

7.1 Summary

This thesis deals with the design of a higher order mode beam position monitor (HOMBPM) using dipole modes in third harmonic 3.9 GHz accelerating cavities. HOMBPM has the capability of measuring beam positions inside the accelerating module without additional vacuum components. Conventional BPMs fall short of providing this information directly. The amplitude of the dipole mode has a linear dependence on the transverse position of the excitation bunch. This enables the diagnostics by measuring these dipole modes extracted from HOM couplers.

Although the principle of beam diagnostics with HOM has been proved in the TESLA 1.3 GHz cavity at FLASH [1], the realization of a HOMBPM in 3.9 GHz cavities is considerably more challenging due to the complexity present in the HOM spectrum. The first obstacle is to find suitable dipole modes. The ideal candidate is the mode which has strong coupling to the beam (large R/Q value), since it can deliver higher position resolution. In 3.9 GHz cavities, the first two dipole bands contain modes with large R/Q's. However, these modes are not confined inside each cavity but propagating through the entire four-cavity module. HOMBPM using these multi-cavity modes can only deliver the beam position in a four-cavity module sense. The second obstacle is that modal frequencies are unpredictable. This inevitably adds complexities to the electronics design for each HOM coupler. Some modes in the fifth dipole band are found to be trapped inside each cavity and could be used for local position diagnostics. The third obstacle is that these modes are located in an upper frequency range ($\sim$9 GHz) and have only weak coupling to the beam (small R/Q values). The former will require careful electronics design and the latter will impact the position resolution of the HOMBPM.

In order to overcome or compromise these obstacles, extensive studies have been conducted prior to building the HOMBPM electronics. These are eigenmode simulations (Chap. 2), measurements of the scattering parameter S_{21} (Chap. 3), beam-based HOM spectra measurements (Chap. 3) and the study of dipole dependence on the beam offset with a spectrum analyzer (Chap. 5) [2]. Three modal options promising

P. Zhang, *Beam Diagnostics in Superconducting Accelerating Cavities*, Springer Theses, DOI: 10.1007/978-3-319-00759-5_7,

for beam diagnostics have been narrowed down: localized dipole beam-pipe modes at approximately 4 GHz, coupled cavities modes in the first two dipole bands in the range of 4.5–5.5 GHz and trapped cavity modes in the fifth dipole band at approximately 9 GHz. In order to evaluate the position resolution using these options, a test electronics consisting of a downmixer and digitizer was designed and built by Fermilab. Various frequency bands of these three options have been studied with the test electronics (Chap. 6) [3]. In the end, two types of HOMBPM appear to be feasible. The first type is based on a four-cavity module using coupled cavity modes with large R/Q's in the first two dipole bands. Since the 3.9 GHz four-cavity string is not significantly longer than one TESLA 1.3 GHz cavity (approximately 1.3 times), the beam position determined in a module sense is still useful. In addition, coupled strongly to the beam, these modes are essential for beam alignment to minimize HOM power. The second type is based on each cavity using localized cavity modes in the fifth dipole band. Moreover, we utilize the signal of a complete band of frequencies rather than a specific dipole mode as used in HOMBPM system for TELSA 1.3 GHz cavities. This has been proved to be able to not only improve the resolution previously limited by modes with small R/Q's, but also accommodate the individual modal frequencies among couplers.

Various data analysis methods have been applied to effectively correlate HOMs to the beam offset (Chap. 4) [4]. A Direct Linear Regression (DLR) on the HOM signals is able to predict beam positions accurately but fell short of computation efficiency due to a considerable number of sampling points of the HOM signals. Thus two statistical methods have been applied and successfully reduced the dimension of the HOM signal without losing the concealed beam position information: Singular Value Decomposition (SVD) and k-means Clustering. Then both accurate position predictions and computation efficiencies have been achieved along with a noise reduction as a byproduct.

For the case of only one bunch in the train, our present experiments with the test electronics suggest a resolution of 50 μ m accuracy in predicting local beam position in the cavity and a global resolution of 20 μ m over the complete module. For multi-bunch operation, a non-degenerate resolution can be achieved when diagnosing the first bunch in the train by using a proper time window on the HOM signals. Due to an overlap of HOM signals with the following bunches, degeneration of position resolution can be foreseen for the trailing bunches in the train (Chap. 6).

There are three highlights present in this work. This is the first time that dependencies of HOMs on beam offset have been observed in third harmonic 3.9 GHz cavities. A new analysis method named k-means Clustering has been successfully implemented in the HOM-based beam diagnostics for the first time. This is also the first time that three data analysis techniques (DLR, SVD and k-means Clustering) have been compared directly in terms of performance and model complexity in the study of beam diagnostics with HOMs.

7.2 Future Studies on HOMBPM

Based on the extensive evaluations of various frequency bands using the test electronics, a set of dedicated electronics is being built using modes in the second dipole band and the fifth dipole band, in order to facilitate beam position measurements over the whole third harmonic module with high resolution, and localized within individual 3.9 GHz cavities with reasonable resolution. The center frequency has been chosen to be 5,440 MHz for the coupled modes and 9,060 MHz for the trapped cavity modes both with 20 MHz bandwidth. For FLASH, we will equip two HOM couplers for the 5,440 MHz and six couplers for the 9,060 MHz except the first and the last HOM couplers in the module. We plan to evaluate the HOMBPM system and integrated into FLASH control system in 2013.

As observed in the previous TESLA 1.3 GHz cavity case, the calibration of the HOMBPM was only stable for 1–2 days. This issue need to be understood and is planned to be studied for both 1.3 and 3.9 GHz HOMBPM system. This is essential for a stable operation of an online beam diagnostics instrument. Based on the results obtained so far, it has been planned to build similar HOMBPM system for the European XFEL third harmonic module consisting of eight 3.9 GHz cavities.

References

1. S. Molloy et al., High precision superconducting cavity diagnostics with higher order mode measurements. Phys. Rev. ST Accel. Beams **9**, 112801 (2006)
2. P. Zhang, N. Baboi, R.M. Jones, I.R.R. Shinton, T. Flisgen, H.W. Glock, A study of beam position diagnostics using beam-excited dipole modes in third harmonic superconducting accelerating cavities at a free-electron laser. Rev. Sci. Instrum. **83**, 085117 (2012)
3. P. Zhang, N. Baboi, R.M. Jones, N. Eddy, Resolution study of higher-order-mode-based beam position diagnostics using custom-built electronics in strongly coupled 3.9 GHz multi-cavity accelerating module. J. Instrum. **7**, P11016 (2012)
4. P. Zhang, N. Baboi and R.M. Jones, Statistical methods for transverse beam position diagnostics with higher order modes in third harmonic 3.9 GHz superconducting accelerating cavities at FLASH. Nucl. Instr. Meth. Phys. Res. (2012). http://dx.doi.org/10.1016/j.nima.2012.11.057

Appendix A
Mathematics

A.1 Direct Linear Regression

To define the problem of linear regression, we have the feature vector $x \in \mathbb{R}^n$ and the target vector $y \in \mathbb{R}^1$. Assuming that we have a total number of m events,s and the ith event is then denoted as $(x^{(i)}, y^{(i)})$, while the jth feature of the ith event is then expressed as $x_j^{(i)}$. In a linear regression approach, we search for the prediction of the target y (hypotheses h) as a linear function of x,

$$h(x) = \sum_{j=0}^{n} \theta_i x_i, \tag{A.1}$$

where θ_i is the ith coefficient and n is the number of features in each event. The task of linear regression is to find the coefficients θ by making $h(x)$ close to y. To formalize this, we define a function that measures, for each value of the θ's, how close the $h(x^{(i)})$'s are to the corresponding $y^{(i)}$'s. We define the *cost function*:

$$J(\theta) = \frac{1}{2} \sum_{i=1}^{m} (h_\theta(x^{(i)}) - y^{(i)})^2. \tag{A.2}$$

This is a typical least-squares cost function for the ordinary least squares regression model. The goal is to minimize $J(\theta)$ by optimizing θ. It starts with some initial guess for θ, and then repeatedly changes θ to make $J(\theta)$ smaller, until we converge to a value of θ that minimizes $J(\theta)$. The algorithm used to perform the ordinary least squares regression is called *gradient descent* [1]. It starts with some initial θ, and repeatedly performs the update:

$$\theta_j \rightarrow \theta_j - \alpha \frac{\partial}{\partial \theta_j} J(\theta), \tag{A.3}$$

P. Zhang, *Beam Diagnostics in Superconducting Accelerating Cavities*,
Springer Theses, DOI: 10.1007/978-3-319-00759-5,

where α is called the learning rate, which controls the changing step of θ_j. This update is simultaneously performed for all values of $j = 0, \ldots, n$. This is a very natural algorithm that repeatedly takes a step in the direction of steepest decrease of J. In order to implement this algorithm, the partial derivative term on the right hand side is derived by the Widrow-Hoff learning rule. The algorithm then becomes

Repeat until convergence {

$$\theta_j \rightarrow \theta_j + \alpha \sum_{i=1}^{m} (y^{(i)} - h_\theta(x^{(i)}))x_j^{(i)} \text{(for every } j\text{)}. \tag{A.4}$$

}

This rule has several properties that is natural and intuitive. For instance, the magnitude of the update is proportional to the error term $(y^{(i)} - h_\theta(x^{(i)}))$; thus, for instance, if we are encountering a training example on which our prediction nearly matches the actual value of $y^{(i)}$, then we find that there is little need to change the parameters; in contrast, a larger change to the parameters will be made if our prediction $h_\theta(x^{(i)})$ has a large error (i.e., if it is very far from $y^{(i)}$). One need to note that this method looks at every event in the entire dataset on every step, and is often called *batch gradient descent*. This is a costly method if m is large, and there is an alternative algorithm called *stochastic gradient descent*, which update the parameters θ according to the gradient of the error with respect to that single data event only [2].

A.2 Singular Value Decomposition

Theorem A.2.1 An $m \times n$ real[1] matrix $A \in \mathbb{R}^{m \times n}$ can be factorized using singular value decomposition as

$$A = U \cdot S \cdot V^T, \tag{A.5}$$

$$U = [u_1, \ldots, u_m] \in \mathbb{R}^{m \times m}, \tag{A.6}$$

$$V^T = [v_1, \ldots, v_n] \in \mathbb{R}^{n \times n}, \tag{A.7}$$

$$S = diag(\sigma_1, \ldots, \sigma_p), \ p = min(m, n) \tag{A.8}$$

where

$$\sigma_1 \geqslant \sigma_2 \geqslant \cdots \geqslant \sigma_p \geqslant 0. \tag{A.9}$$

U is a $m \times m$ unitary matrix, S is a $m \times n$ rectangular diagonal matrix with nonnegative real numbers on the diagonal (the first p elements on the diagonal are non-zero), V^T is a $n \times n$ unitary matrix and is the transpose of V. The proof of

[1] The complex matrix A can also be factorized by SVD and is discussed in [3].

the theorem can be found in [3]. The σ_i are the singular values of A. The vector u_i and v_i are the ith left singular vector and the ith right singular vector respectively. Since U and V are unitary, the columns of U and V, u_i's and v_i's, form two sets of orthonormal vectors, which can be regarded as basis vectors. From the SVD theorem, a useful corollary is listed below and the proof can be found in [3].

Corollary A.2.1 Let the SVD of $A \in \mathbb{R}^{m\times n}$ be given by Theorem A.2.1, and rank(A)$= r$. If $k < r$ and

$$A_k = \sum_{i=1}^{k} \sigma_i u_i v_i^T, \tag{A.10}$$

then the 2-norm distance[2]

$$\| A - A_k \| = \sigma_{k+1}. \tag{A.11}$$

Corollary A.2.1 says that the 2-norm difference between the rank k approximation and the original matrix is equivalent to the k+1 singular value.

A.3 Definitions of r^2 and $\bar{r}^2$

Once the prediction coefficients are found, we need to measure how well the predictions are compared to the actual target value y. As defined in Appendix A.1, we have a total number of m events. The target value from the measurement is denoted as y_i, then the mean value of y_i can be calculated as,

$$\bar{y} = \frac{\sum_{i=1}^{m} y_i}{m}. \tag{A.12}$$

The prediction of the target from the regression model is denoted as $\hat{y}_i$. The deviations of actual values from predicted values are prediction errors, and is called the residuals. Suppose that we compare two prediction methods: one using the regression model, the other ignoring the model and always predicting the mean y value. In the first method using regression model, the prediction error can be summarized by the sum of squared residuals (due to errors),

$$SSE = \sum_{i=1}^{m} (y_i - \hat{y}_i)^2. \tag{A.13}$$

In the second method, if we do not use any model, the best squared error predictor of y would be the mean value $\bar{y}$. Then the sum of squared prediction errors would

[2] The 2-norm distance is defined as the rms sum of the differences of each of the matrix elements [3].

be, in this case, about the mean,

$$SSM = \sum_{i=1}^{m} (y_i - \bar{y})^2. \tag{A.14}$$

For comparison, the proportionate reduction in error can be defined and is called *coefficient of determination* [4]:

$$r^2 = \frac{SSM - SSE}{SSM} = 1 - \frac{\sum_{i=1}^{m}(y_i - \hat{y}_i)^2}{\sum_{i=1}^{m}(y_i - \bar{y})^2}. \tag{A.15}$$

That means, use of the regression model reduces the squared prediction error by r^2. This is a measurement of how strong the relation between features and targets by using our regression model. From the definition, the value of r^2 has the range from 0 to 1. A r^2 of zero indicates no predictive power of the regression model, in other words, one can predict y without knowing x as well as one can by knowing x. A r^2 of one indicates perfect predictability, meaning that 100 % reduction in error due to the knowledge of x.

There is a limitation in using r^2 to compare different regression models. In general, r^2 will eventually be very close to one as more regressors are included into the model, in other words, more complex model tends to have larger r^2. Thus, r^2 can give misleading results if we are trying to balance the two criteria of obtaining a model in which we have a good fit and a model in which we have a limited number of regressors. Therefore, we define an adjusted r^2 [5], which provides a penalty for each regression coefficient included in the model:

$$r_{adj}^2 = 1 - \frac{SSE/(n-p-1)}{SSM/(n-1)} = 1 - (1 - r^2)\frac{n-1}{n-p-1}, \tag{A.16}$$

where n is the total number of the regressors which can be included in the model, p is the number of regressors currently included in the model. Note that in r_{adj}^2, the sum of squares are adjusted for their corresponding degrees of freedom. Also, increasing the number of regressors in the model from p to $p + 1$ will not always result in an increase in r_{adj}^2, as would be true for r^2. If including the new regressor can not improve the fit of the model to the data by a reasonable level, then the performance will be penalized with a smaller r_{adj}^2 value.

A.4 Definitions of Resolution

Resolution is the smallest change that can be distinguished by the device. In the calibration and validation stage, we define the resolution as the rms (root mean square) of the differences between measurements and predictions. Assume that we have measured value y_i and predicted value $\hat{y}_i$. Then the resolution for y is:

$$Resolution = \sqrt{\frac{\sum_{i=1}^{m}(y_i - \hat{y}_i)^2}{m}} = \sqrt{\frac{SSE}{m}}. \tag{A.17}$$

References

1. J.A. Snyman, *Practical Mathematical Optimization: An Introduction to Basic Optimization Theory and Classical and New Gradient-Based Algorithms*, ch. 4.3.2. (Springer Publishing, 2005), p. 106
2. C.M. Bishop, *Pattern Recognition and Machine Learning*, ch. 3.1.3, 1st edn. (Springer Publishing, 2006), pp. 143–144
3. G.H. Golub, C.F. Van Loan, *Matrix Computations*, ch. 2.3, 2nd edn. (The John Hopkins University Press, 1984), pp. 16–20
4. R.L. Ott, M. Longnecker, *An Introduction to Statistical Methods and Data Analysis*, ch. 11.7, 6th edn. (Duxbury Press, 2008), p. 611
5. R.L. Ott, M. Longnecker, *An Introduction to Statistical Methods and Data Analysis*, ch. 13.2, 6th edn. (Duxbury Press, 2008), p. 771

Appendix B
Eigenmodes of an Ideal Third Harmonic Cavity

B.1 List of Eigenmodes

The frequencies and R/Q's of eigenmodes simulated on an ideal 9-cell third harmonic cavity are listed in this section. The modes are grouped in bands: "M" denotes monopole, "M1" denotes the first monopole band, "M1-1" denotes the first mode in M1, "D" denotes dipole, "Q" denotes quadrupole and "S" denotes sextupole. Beam-pipe modes in each table are denoted as "BP".

B.2 Electric Field Distributions

The electric field distributions of selected modes are presented along with key parameters set in CST Microwave studio® during simulations. In the"Mesh Type" column, "PBA" denotes "perfect boundary approximation" while "FPBA" denotes "fast PBA" with "Enhanced PBA accuracy". The choice between "PBA" and "FPBA" is based on one principle in this study: "FPBA" was used only if "PBA" failed to generate a valid mesh. In the "Mesh cells (million)" column, the number of mesh cells is for a quarter of the structure. The naming in the "Band" column follows the convention explained in Appendix B.1. All modes simulated up to 10 GHz can be found in Ref. [1].

P. Zhang, *Beam Diagnostics in Superconducting Accelerating Cavities*,
Springer Theses, DOI: 10.1007/978-3-319-00759-5,

Table B.1 Monopole modes with electric (EE) or magnetic (MM) boundaries (part 1)

	EE		MM	
Band	f(GHz)	R/Q(Ω)	f(GHz)	R/Q(Ω)
M1-1	3.7466	0.008	3.7466	0.008
M1-2	3.7601	0.061	3.7601	0.061
M1-3	3.7808	0.090	3.7808	0.090
M1-4	3.8065	0.170	3.8065	0.170
M1-5	3.8340	0.307	3.8341	0.309
M1-6	3.8602	0.203	3.8602	0.204
M1-7	3.8817	0.468	3.8817	0.481
M1-8	3.8958	0.195	3.8958	0.197
M1-9	3.9008	373.113	3.9008	373.097
BP1-1	5.7685	6.967	5.8624	0.051
BP1-2	5.7685	2.443	5.8624	0.051
BP2-1	6.0123	2.867	6.2095	0.879
BP2-2	6.0123	3.335	6.2095	0.665
BP3-1	6.4403	5.212	6.6886	4.119
BP3-2	6.4403	6.116	6.6886	2.302
BP4-1	6.9393	8.454		
BP4-2	6.9394	3.971		
M2-1	7.0483	0.127	7.0449	0.007
M2-2	7.0863	1.012	7.0738	0.034
M2-3	7.1424	0.088	7.1145	0.342
M2-4	7.2113	2.914	7.1589	0.224
M2-5	7.2877	0.677	7.2058	3.534
M2-6	7.3662	2.727	7.2625	4.008
M2-7	7.4418	10.686	7.3331	2.417
M2-8	7.5118	18.963	7.4140	14.981
M2-9	7.5843	47.909	7.4995	28.455
M2-10			7.5810	3.471

Table B.2 Monopole modes with electric (EE) or magnetic (MM) boundaries (part 2)

	EE		MM	
Band	f(GHz)	R/Q(Ω)	f(GHz)	R/Q(Ω)
M3-1	7.6443	0.593	7.6780	54.835
M3-2	7.7248	21.212	7.7577	0.095
M3-3	7.8036	0.245	7.8333	0.619
M3-4	7.8809	1.707	7.9021	0.000
M3-5	7.9547	0.640	7.9625	0.275
M3-6	8.0229	0.015	8.0148	0.756
M3-7	8.0829	0.113	8.0615	0.117
M3-8	8.1311	0.014	8.1044	0.755
M3-9	8.1631	0.001	8.1408	0.002
M3-10			8.1656	0.078
BP5-1	8.3376	0.765	8.7111	0.521
BP5-2	8.3376	1.045	8.7113	0.525
BP6-1	9.1202	1.670	9.5377	1.912
BP6-2	9.1202	1.755	9.5378	1.945
M4-1	9.7966	0.000	9.7907	0.000
M4-2	9.8340	1.511	9.8379	0.072
M4-3	9.8868	0.395	9.9124	0.360
M4-4	9.9384	7.388	10.0099	5.067
M4-5	9.9886	0.015	10.1270	0.264
M4-6	10.0619	0.921	10.2547	0.723
M4-7	10.1692	1.125	10.3692	2.342
M4-8	10.3015	0.778	10.4406	0.312
M4-9	10.4485	0.277	10.5080	0.719

Table B.3 Dipole modes with electric (EE) or magnetic (MM) boundaries (part 1)

	EE		MM	
Band	f(GHz)	R/Q(Ω/cm^2)	f(GHz)	R/Q(Ω/cm^2)
BP1-1	4.1489	0.234	4.1474	0.241
BP1-2	4.1491	1.318	4.1475	1.299
D1-1	4.2982	0.001	4.2979	0.001
D1-2	4.3607	0.292	4.3592	0.263
D1-3	4.4485	0.002	4.4306	0.072
D1-4	4.5410	1.076	4.4516	0.000
D1-5	4.5989	0.784	4.4770	0.327
D1-6	4.6415	0.165	4.5703	1.213
D1-7	4.7245	10.572	4.6804	1.586
D1-8	4.8327	50.307	4.7749	27.165
D1-9	4.9270	30.174	4.8455	32.124
D1-10	4.9899	0.000	4.9162	21.833
BP2-1	5.2014	0.300	4.9945	7.376
BP2-2	5.2040	2.036	5.0233	3.844
D2-1	5.3581	0.041	5.3518	0.055
D2-2	5.4050	5.057	5.3923	2.114
D2-3	5.4427	20.877	5.4272	10.770
D2-4	5.4678	15.776	5.4528	17.024
D2-5	5.4829	0.895	5.4711	9.368
D2-6	5.4911	1.261	5.4834	0.409
D2-7	5.4950	0.307	5.4908	0.343
D2-8	5.4958	0.549	5.4944	0.033
BP3-1	5.8644	1.028	5.5532	2.994
BP3-2	5.8644	1.026	5.5532	2.995
BP4-1	6.5593	0.344	6.2123	0.595
BP4-2	6.5594	0.397	6.2144	0.636
D3-1	6.8238	0.011	6.7964	0.068
D3-2	6.9003	0.035	6.8242	0.068
D3-3	7.0027	0.058	6.8909	0.140
D3-4	7.1225	0.189	6.9880	0.124
D3-5	7.2541	0.549	7.0989	0.108
D3-6	7.3833	0.014	7.2140	0.020
D3-7	7.4889	0.455	7.3348	0.825
D3-8	7.5621	0.269	7.4598	0.503
D3-9	7.6196	1.354	7.5743	2.862
D3-10	7.6680	28.926	7.6566	23.875
BP5-1	8.3033	1.543	7.9033	2.155
BP5-2	8.3039	1.537	7.9034	2.961

Table B.4 Dipole modes with electric (EE) or magnetic (MM) boundaries (part 2)

	EE		MM	
Band	f(GHz)	R/Q(Ω/cm^2)	f(GHz)	R/Q(Ω/cm^2)
D4-1	8.5002	0.130	8.5292	0.365
D4-2	8.5042	0.096	8.5709	0.023
D4-3	8.5322	0.152	8.6273	1.457
D4-4	8.5763	0.115	8.6849	0.042
D4-5	8.6397	0.415	8.7257	4.392
D4-6	8.7205	1.038	8.7702	1.693
D4-7	8.8033	10.205	8.8301	5.577
D4-8	8.8648	2.470	8.8729	1.895
D4-9	8.9196	0.287	8.9196	0.281
D4-10	8.9857	3.258	8.9900	3.623
D4-11	8.9980	0.230	9.0011	0.302
D5-1	9.0523	0.002	9.0593	0.004
D5-2	9.0530	0.053	9.0599	0.058
D5-3	9.0546	0.058	9.0614	0.076
D5-4	9.0581	2.171	9.0645	2.377
D5-5	9.0664	4.116	9.0718	4.158
D5-6	9.0890	0.580	9.0918	0.452
BP6-1	9.1666	1.291	8.4964	0.340
BP6-2	9.1678	1.898	8.5008	0.056
BP7-1	9.1749	1.123	9.2324	0.158
BP7-2	9.1763	3.240	9.2325	0.092
BP8-1	9.3283	0.880	9.4325	0.105
BP8-2	9.3284	2.042	9.4330	0.020
BP9-1	9.5379	4.064	9.5809	0.342
BP9-2	9.5385	0.275	9.5818	0.015
BP10-1			9.6342	0.485
BP10-2			9.6346	0.848
D6-1	9.6962	0.001	9.6896	0.013
D6-2	9.7142	0.009	9.7103	0.025
D6-3	9.7421	0.074	9.7415	0.341
D6-4	9.7711	0.379	9.7776	1.133
D6-5	9.7896	5.951	9.8134	0.345
D6-6	9.8027	0.771	9.8440	0.008
D6-7	9.8265	0.191	9.8648	0.019

Table B.5 Quadrupole modes with electric (EE) or magnetic (MM) boundaries

	EE		MM	
Band	f(GHz)	R/Q(Ω/cm^4)	f(GHz)	R/Q(Ω/cm^4)
BP1-1	6.2697	0.513	6.2697	0.513
BP1-2	6.2698	0.742	6.2697	0.742
Q1-1	6.5638	0.183	6.5638	0.183
Q1-2	6.5843	3.734	6.5843	3.734
Q1-3	6.6167	4.358	6.6167	4.359
Q1-4	6.6583	0.183	6.6583	0.183
Q1-5	6.7059	0.308	6.7059	0.307
Q1-6	6.7546	0.002	6.7546	0.002
Q1-7	6.7961	0.041	6.7961	0.041
Q2-1	7.0005	0.135	7.0005	0.135
Q2-2	7.0096	0.075	7.0096	0.075
Q2-3	7.0456	0.152	7.0456	0.151
Q2-4	7.0823	0.000	7.0823	0.000
Q2-5	7.1158	0.221	7.1157	0.220
Q2-6	7.1437	0.101	7.1436	0.103
Q2-7	7.1653	0.579	7.1653	0.578
Q2-8	7.1806	3.484	7.1806	3.483
Q2-9	7.1897	2.125	7.1897	2.125
BP2-1	7.4084	0.020	7.3248	0.008
BP2-2	7.4085	0.020	7.3249	0.008
BP3-1	7.7101	0.049	7.5397	0.030
BP3-2	7.7102	0.033	7.5398	0.030
BP4-1	8.1423	0.069	7.9134	0.052
BP4-2	8.1431	0.069	7.9136	0.054
BP5-1	8.6665	0.095	8.3961	0.086
BP5-2	8.6665	0.108	8.3964	0.085
Q3-1	9.1129	4.686	9.1147	3.176
Q3-2	9.1133	0.053	9.1155	0.147
Q3-3	9.1228	5.921	9.1232	7.555
Q3-4	9.1340	11.102	9.1344	11.244
Q3-5	9.1499	2.998	9.1501	2.608
Q3-6	9.1692	0.003	9.1692	0.021
Q3-7	9.1893	0.152	9.1890	0.152
Q3-8	9.2053	0.000	9.2048	0.001
BP6-1	9.2980	0.178	8.9709	0.167
BP6-2	9.2980	0.179	8.9711	0.166

Table B.6 Sextupole modes with electric (EE) or magnetic (MM) boundaries

	EE		MM	
Band	f(GHz)	R/Q(Ω/cm^6)	f(GHz)	R/Q(Ω/cm^6)
BP1-1	7.9214	0.069	7.9212	0.060
BP1-2	7.9216	0.069	7.9215	0.068
S1-1	8.1894	0.506	8.1894	0.506
S1-2	8.1940	0.203	8.1940	0.203
S1-3	8.2011	0.006	8.2011	0.006
S1-4	8.2097	0.027	8.2097	0.027
S1-5	8.2184	0.001	8.2184	0.001
S1-6	8.2261	0.005	8.2261	0.005
S1-7	8.2313	0.000	8.2313	0.000
BP2-1	8.7611	0.015	8.7612	0.018
BP2-2	8.7614	0.013	8.7615	0.017
S2-1	8.8029	0.103	8.8029	0.000
S2-2	8.8097	0.001	8.8097	0.001
S2-3	8.8192	0.003	8.8191	0.002
S2-4	8.8295	0.001	8.8294	0.412
S2-5	8.8392	0.024	8.8391	0.028
S2-6	8.8469	0.036	8.8468	0.035
S2-7	8.8519	0.004	8.8519	0.004

B.2.1 Monopole (Electric Boundaries)

Table B.7 Parameters settings for M1 with electric boundary (EE)

Frequency range (GHz)	Lines per wave-length	Mesh type	Number of mesh cells	Max mesh step (mm)	Band
3.70–3.95	70	PBA	2,100,000	1.10	M1

Table B.8 Monopole modes in M1 with electric boundary (EE)

E-field amplitude	f(GHz)	R/Q(Ω)	Band
	3.7466	0.008	M1-1
	3.7601	0.061	M1-2
	3.7808	0.090	M1-3
	3.8065	0.170	M1-4
	3.8340	0.307	M1-5
	3.8602	0.203	M1-6
	3.8817	0.468	M1-7
	3.8958	0.195	M1-8
	3.9008	373.113	M1-9

B.2.2 Monopole (Magnetic Boundaries)

Table B.9 Parameters settings for M1 with magnetic boundary (MM)

Frequency range (GHz)	Lines per wavelength	Mesh type	Number of mesh cells	Max mesh step (mm)	Band
3.7–4.0	70	PBA	2,200,000	1.10	M1

B.2.3 Dipole (Electric Boundaries)

Table B.10 Monopole modes in M1 with magnetic boundary (MM)

E-field amplitude	f(GHz)	R/Q(Ω)	Band
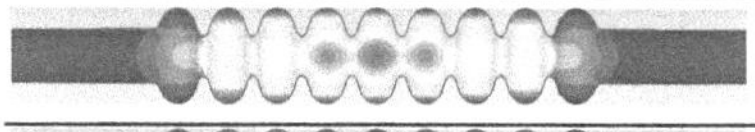	3.7466	0.008	M1-1
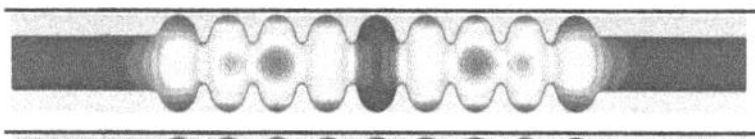	3.7601	0.061	M1-2
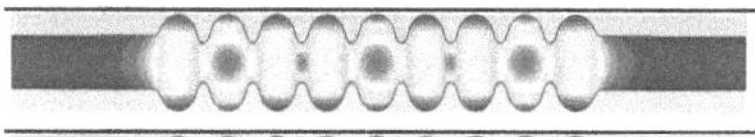	3.7808	0.090	M1-3
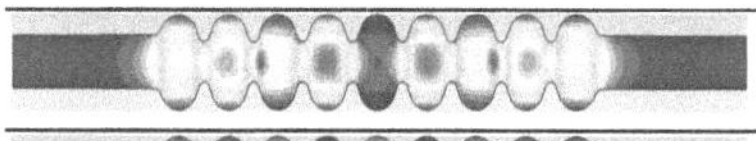	3.8065	0.170	M1-4
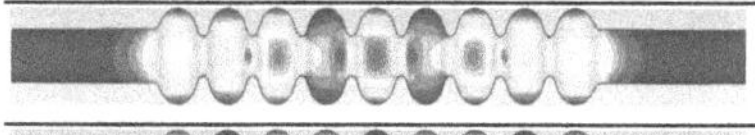	3.8341	0.309	M1-5
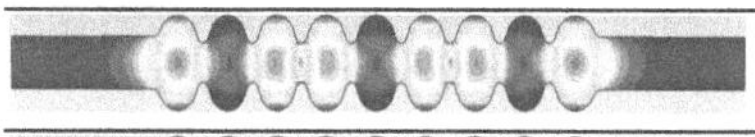	3.8602	0.204	M1-6
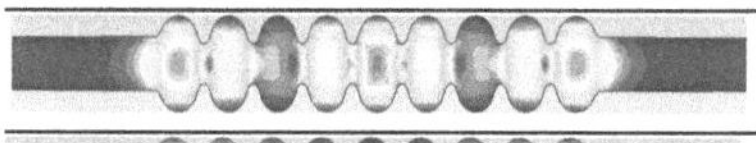	3.8817	0.481	M1-7
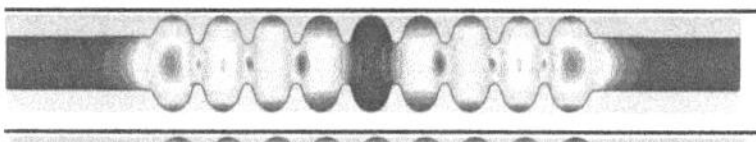	3.8958	0.197	M1-8
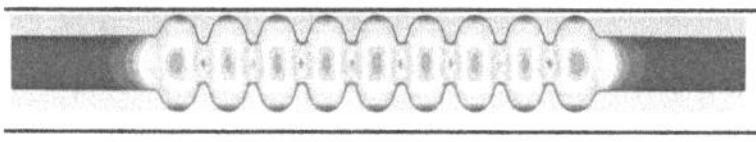	3.9008	373.097	M1-9

B.2.4 Dipole (Magnetic Boundaries)

Table B.11 Parameters settings for DBP1 with electric (EE) boundaries

Frequency range (GHz)	Lines per wave-length	Mesh type	Number of mesh cells	Max mesh step (mm)	Band
4.1–4.2	70	PBA	2,400,000	1.04	DBP1

Table B.12 Dipole modes in DBP1 with electric (EE) boundaries

E-field amplitude	f(GHz)	R/Q(Ω/cm^2)	Band
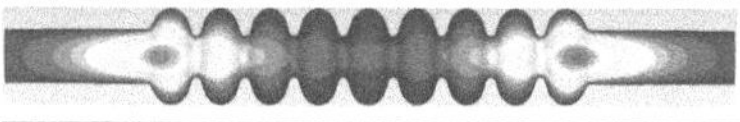	4.1489	0.234	DBP1-1
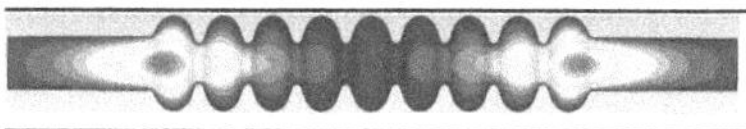	4.1491	1.318	DBP1-2

Table B.13 Parameters settings for D1 with electric (EE) boundaries

Frequency range (GHz)	Lines per wavelength	Mesh type	Number of mesh cells	Max mesh step (mm)	Band
4.25–5.05	60	PBA	2,600,000	1.03	D1(1–9)
4.95–5.30	45	PBA	1,500,000	1.29	D1(10)

Table B.14 Dipole modes in D1 with electric (EE) boundaries

E-field amplitude	f(GHz)	R/Q(Ω/cm^2)	Band
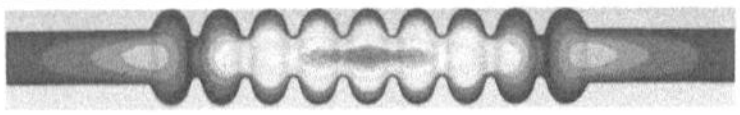	4.2982	0.001	D1-1
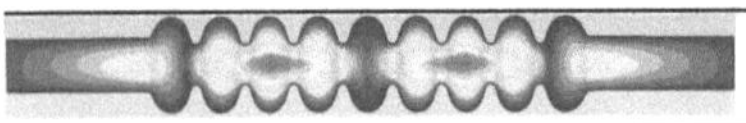	4.3607	0.292	D1-2
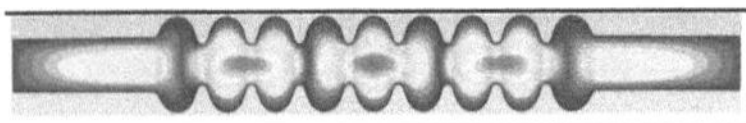	4.4485	0.002	D1-3
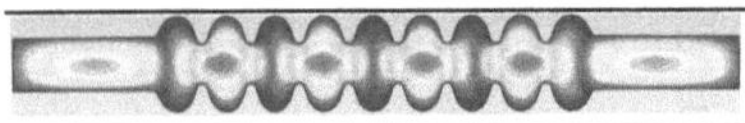	4.5410	1.076	D1-4
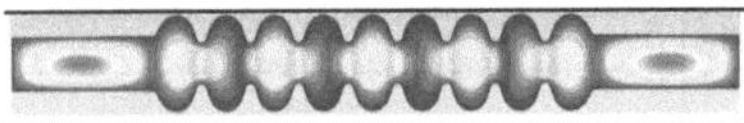	4.5989	0.784	D1-5
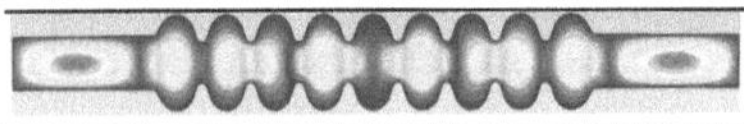	4.6415	0.165	D1-6
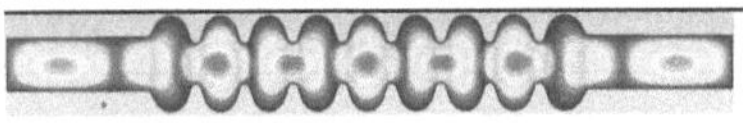	4.7245	10.572	D1-7
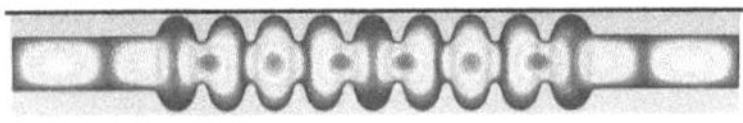	4.8327	50.307	D1-8
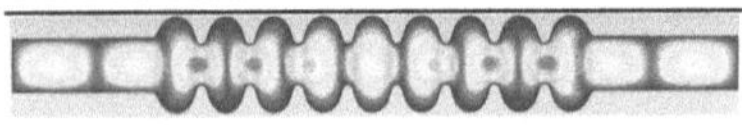	4.9270	30.174	D1-9
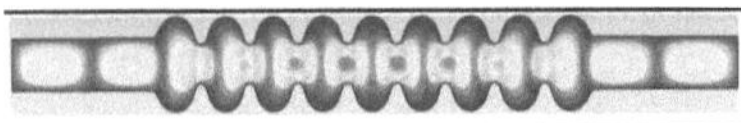	4.9899	0.000	D1-10

Table B.15 Parameters settings for DBP2 and D2 with electric (EE) boundaries

Frequency range (GHz)	Lines per wavelength	Mesh type	Number of mesh cells	Max mesh step (mm)	Band
4.95–5.30	45	PBA	1,500,000	1.29	DBP2
5.3–5.6	50	PBA	2,200,000	1.10	D2

Table B.16 Dipole modes in DBP2 and D2 with electric (EE) boundaries

E-field amplitude	f(GHz)	R/Q(Ω/cm^2)	Band
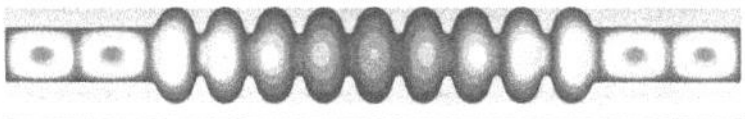	5.2014	0.300	DBP2-1
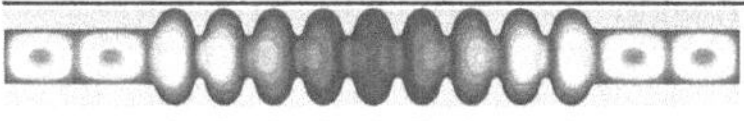	5.2040	2.036	DBP2-2
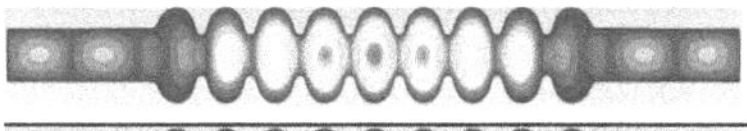	5.3581	0.041	D2-1
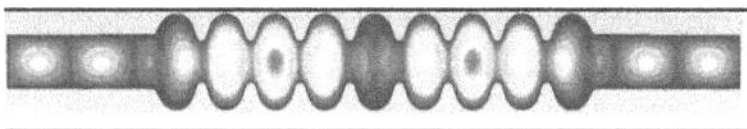	5.4050	5.057	D2-2
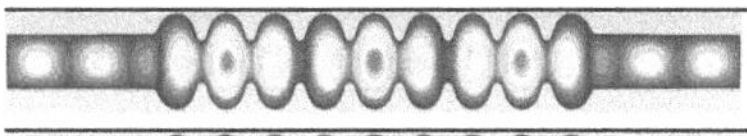	5.4427	20.877	D2-3
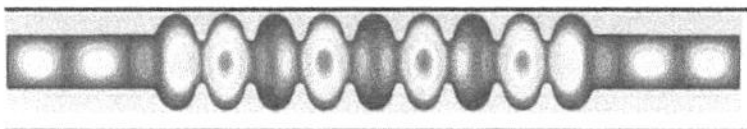	5.4678	15.776	D2-4
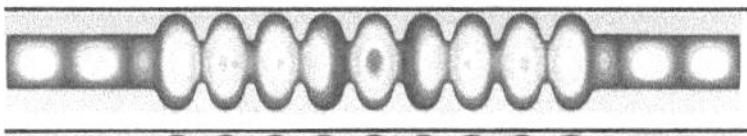	5.4829	0.895	D2-5
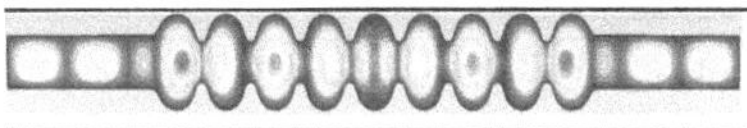	5.4911	1.261	D2-6
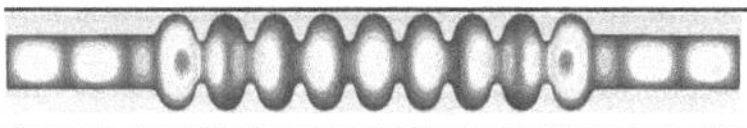	5.4950	0.307	D2-7
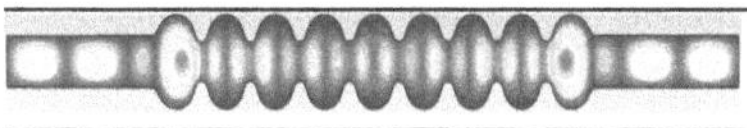	5.4958	0.549	D2-8

Table B.17 Parameters settings for D5 with electric boundary (EE)

Frequency range (GHz)	Lines per wavelength	Mesh type	Number of mesh cells	Max mesh step (mm)	Band
9.05–9.095	20	PBA	1,300,000	0.91	D5

Table B.18 Dipole modes in D5 with electric boundary (EE)

E-field amplitude	f(GHz)	R/Q(Ω/cm^2)	Band
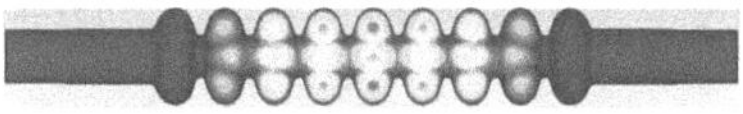	9.0523	0.002	D5-1
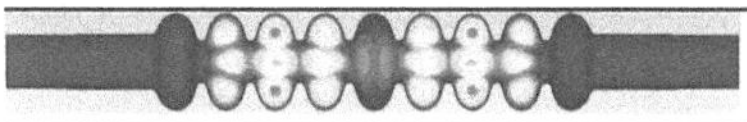	9.0530	0.053	D5-2
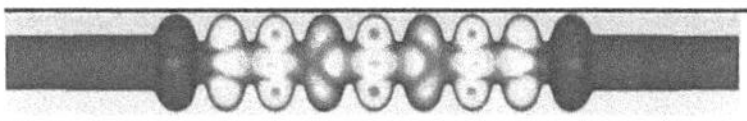	9.0546	0.058	D5-3
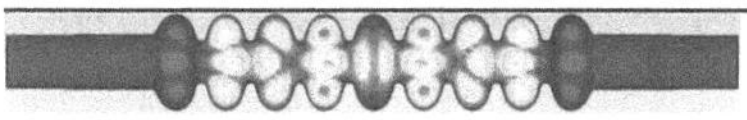	9.0581	2.171	D5-4
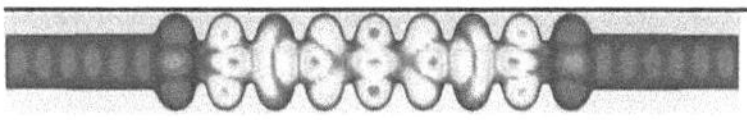	9.0664	4.116	D5-5
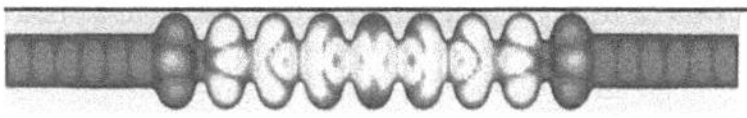	9.0890	0.580	D5-6

Table B.19 Parameters settings for DBP1 with magnetic boundary (MM)

Frequency range (GHz)	Lines per wavelength	Mesh type	Number of mesh cells	Max mesh step (mm)	Band
4.1–4.2	70	PBA	2,400,000	1.04	DBP1

Table B.20 Dipole modes in DBP1 with magnetic boundary (MM)

E-field amplitude	f(GHz)	R/Q(Ω/cm^2)	Band
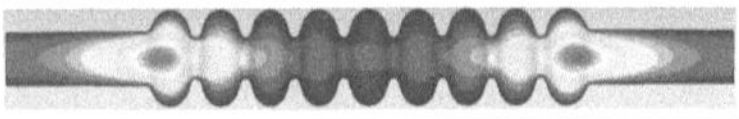	4.1474	0.241	DBP1-1
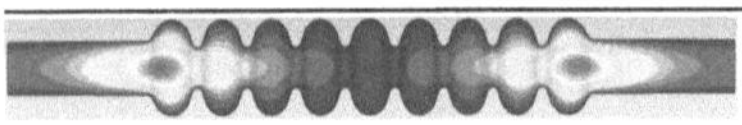	4.1475	1.299	DBP1-2

Table B.21 Parameters settings for D1 with magnetic boundary (MM)

Frequency range (GHz)	Lines per wavelength	Mesh type	Number of mesh cells	Max mesh step (mm)	Band
4.25–5.05	55	PBA	2,100,000	1.10	D1

Table B.22 Dipole modes in D1 with magnetic boundary (MM)

E-field amplitude	f(GHz)	R/Q(Ω/cm^2)	Band
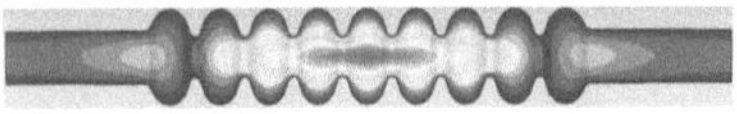	4.2979	0.001	D1-1
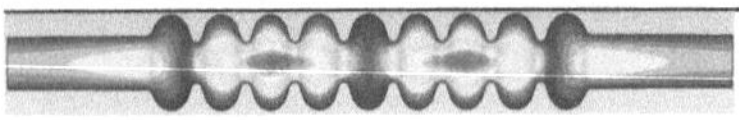	4.3592	0.263	D1-2
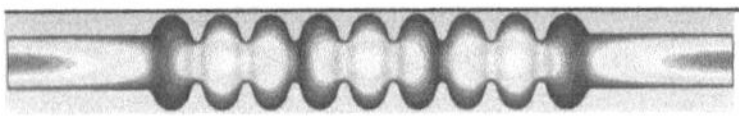	4.4306	0.072	D1-3
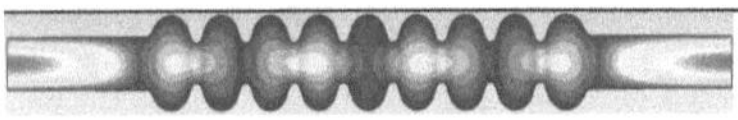	4.4516	0.000	D1-4
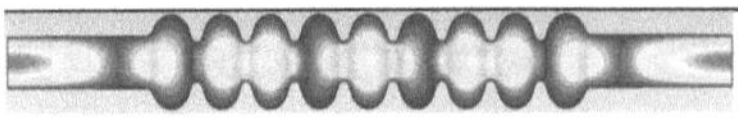	4.4770	0.327	D1-5
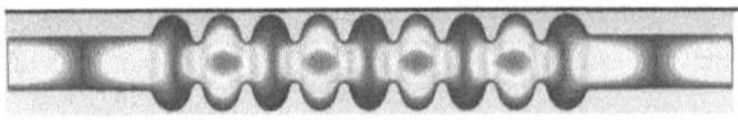	4.5703	1.213	D1-6
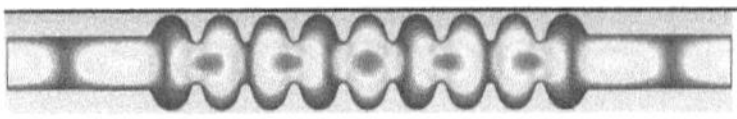	4.6804	1.586	D1-7
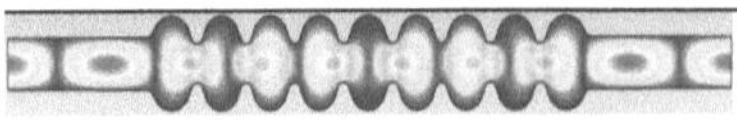	4.7749	27.165	D1-8
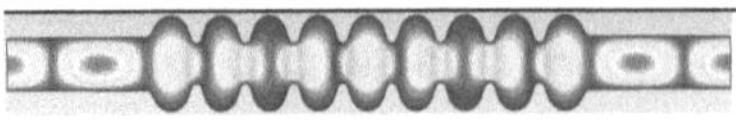	4.8455	32.124	D1-9
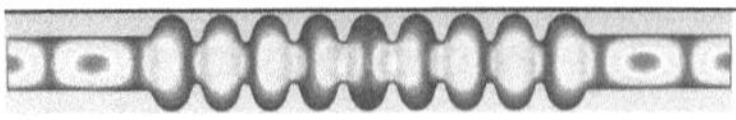	4.9162	21.833	D1-10

Table B.23 Parameters settings for DBP2 and D2 with magnetic boundary (MM)

Frequency range (GHz)	Lines per wavelength	Mesh type	Number of mesh cells	Max mesh step (mm)	Band
4.25–5.05	55	PBA	2,100,000	1.10	DBP2
5.3–5.7	50	PBA	2,300,000	1.07	D2

Table B.24 Dipole modes in DBP2 and D2 with magnetic boundary (MM)

E-field amplitude	f(GHz)	R/Q(Ω/cm²)	Band
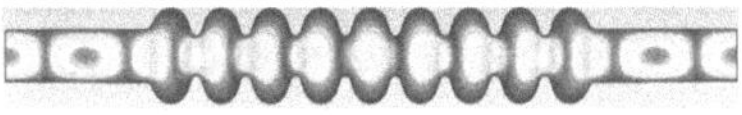	4.9945	7.376	DBP2-1
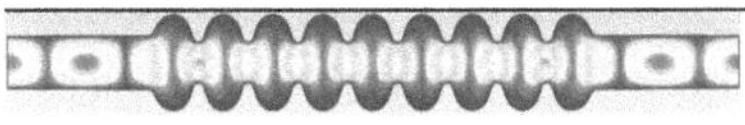	5.0233	3.844	DBP2-2
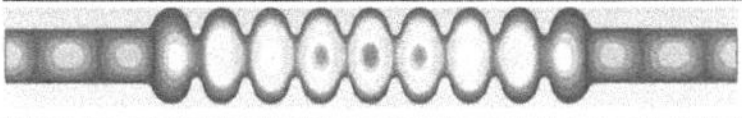	5.3518	0.055	D2-1
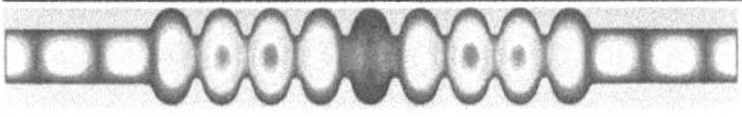	5.3923	2.114	D2-2
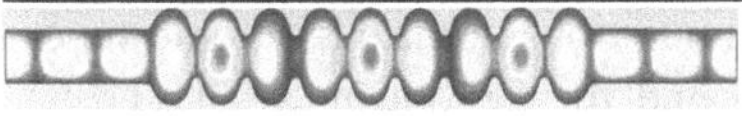	5.4272	10.770	D2-3
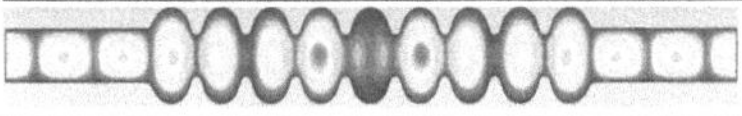	5.4528	17.024	D2-4
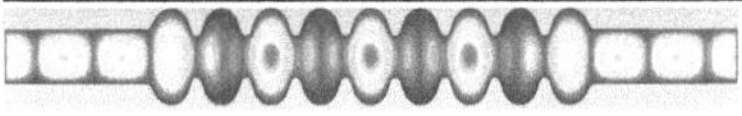	5.4711	9.368	D2-5
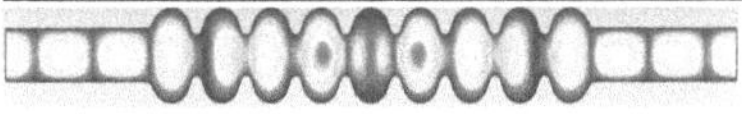	5.4834	0.409	D2-6
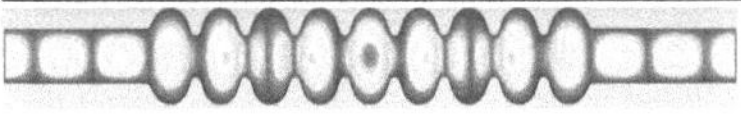	5.4908	0.343	D2-7
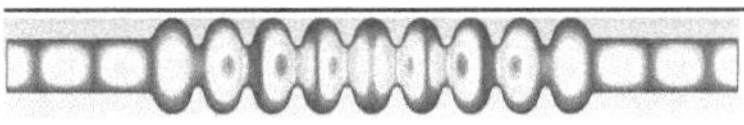	5.4944	0.033	D2-8

Table B.25 Parameters settings for D5 with magnetic boundary (MM)

Frequency range (GHz)	Lines per wavelength	Mesh type	Number of mesh cells	Max mesh step (mm)	Band
9.05–9.10	16	FPBA	800,000	1.08	D5

Table B.26 Dipole modes in D5 with magnetic boundary (MM)

E-field amplitude	f(GHz)	R/Q(Ω/cm²)	Band
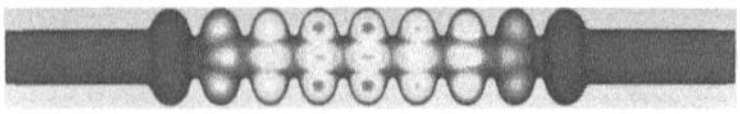	9.0593	0.004	D5-1
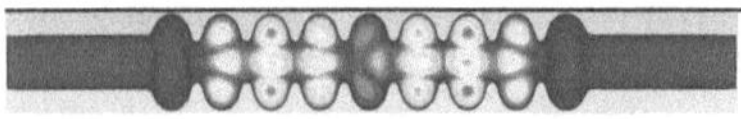	9.0599	0.058	D5-2
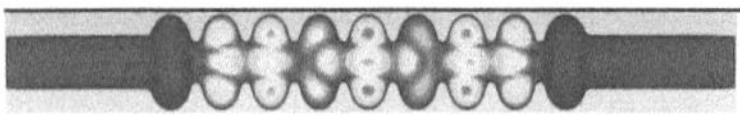	9.0614	0.076	D5-3
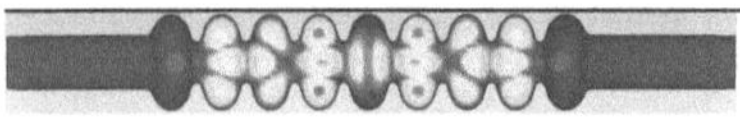	9.0645	2.377	D5-4
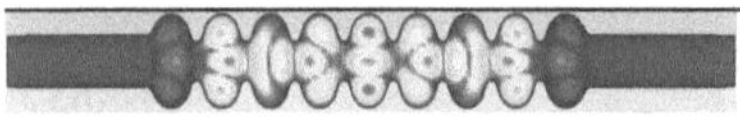	9.0718	4.158	D5-5
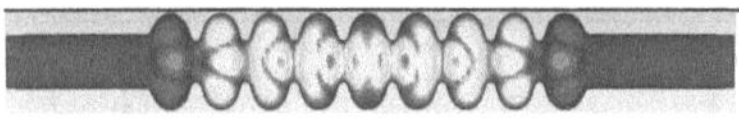	9.0918	0.452	D5-6

References

1. P. Zhang, N. Baboi, R.M. Jones, Eigenmode simulations of third harmonic superconducting accelerating cavities for FLASH and the European XFEL, DESY report: DESY 12-101, 2012

Appendix C
Technical Details of the HOM Measurements

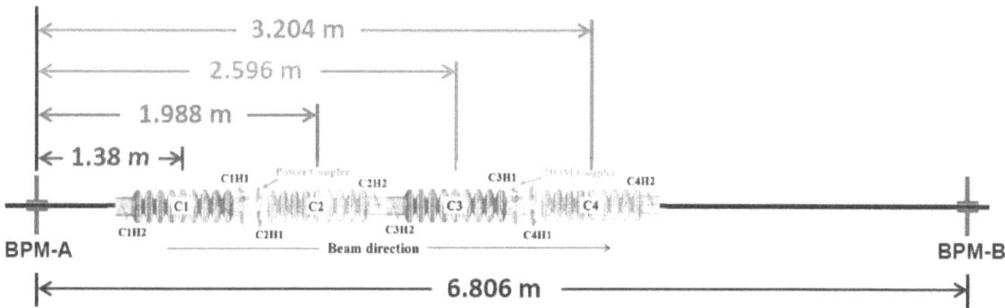

Fig. C.1 Beam position interpolation

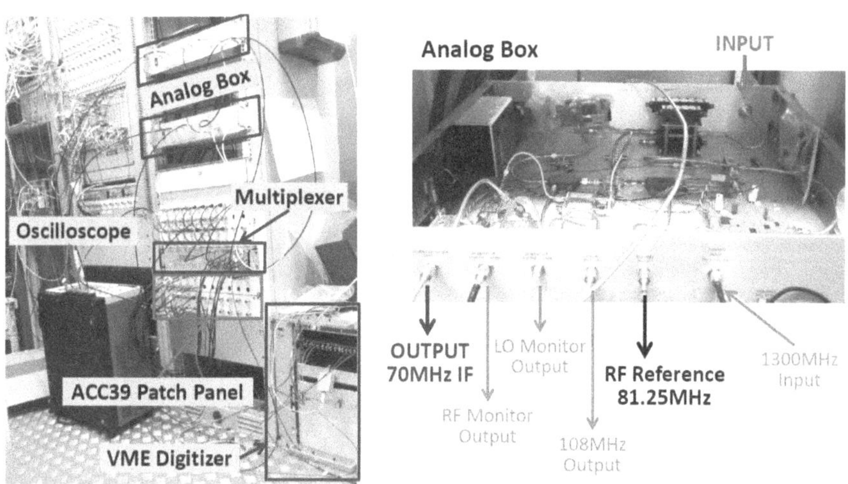

Fig. C.2 Photo of the device setup in the FLASH injector rack outside the tunnel and the analog box

P. Zhang, *Beam Diagnostics in Superconducting Accelerating Cavities*,
Springer Theses, DOI: 10.1007/978-3-319-00759-5,

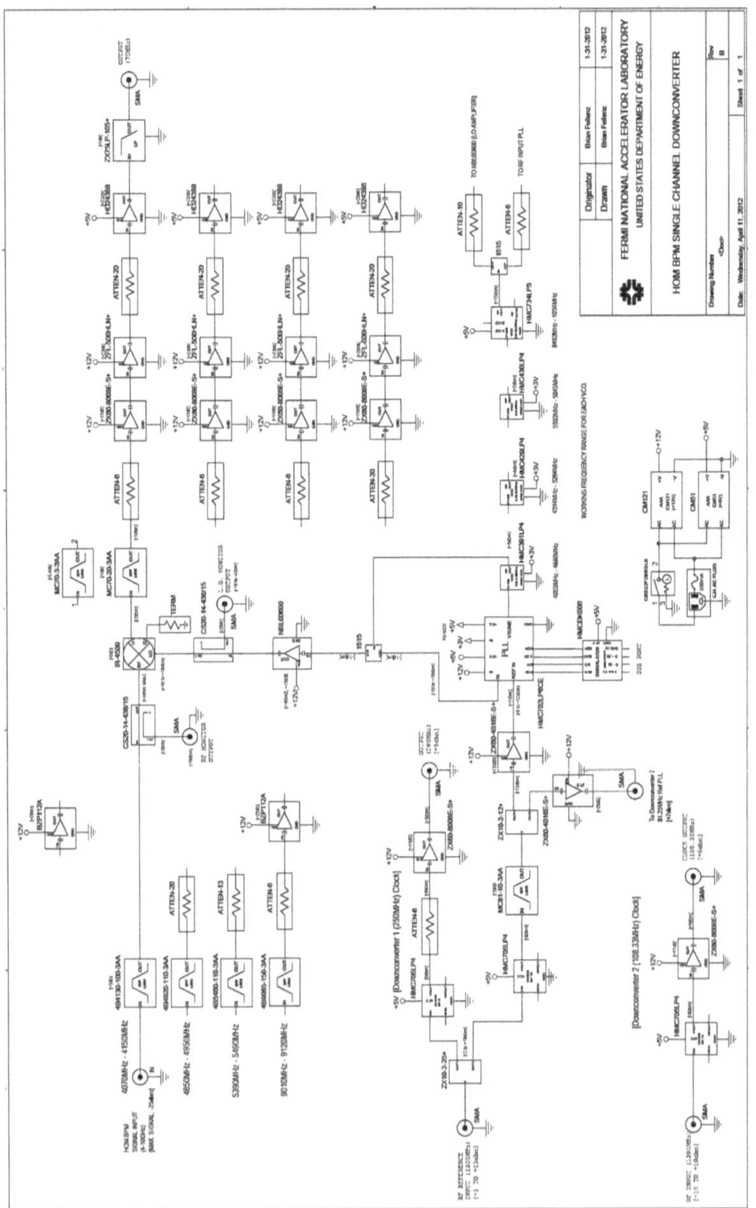

Fig. C.3 Circuit drawing of the analog electronics (Courtesy B. Fellenz and N. Eddy from FNAL)

About the Author

Education

- 2012 Ph.D. in Accelerator Physics, The University of Manchester, UK.
- 2009 M.Sc. in Particle Physics, University of Science and Technology of China (USTC), Hefei, China.
- 2006 B.Sc. in Applied Physics, USTC, Hefei, China.

Appointments Held

- 2013 *Marie Curie Experienced Researcher*, European Organization for Nuclear Research (CERN), Geneva, Switzerland.
- 2010-2012 *Graduate Scientific Assistant*, Deutsches Elektronen Synchrotron (DESY), Hamburg, Germany.
- 2007-2009 *Graduate Scientific Assistant*, USTC, Hefei, China.
- 2006-2007 *Visiting Scholar*, University of Michigan (Ann Arbor, MI, USA) and CERN (Geneva, Switzerland).

Grants, Honors & Awards

- 2011 *European Accelerator Prize*, The European Physical Society - Accelerator Group (EPS-AG).
- 2010 *Distinguished Performance Award*, The Fifth International Accelerator Schools for Linear Colliders.

Publications & Talks
Journal articles

- P. Zhang, N. Baboi, R. M. Jones, I. R. R. Shinton, T. Flisgen and H.-W. Glock, "A study of beam position diagnostics using beam-excited dipole modes in third harmonic superconducting accelerating cavities at a free-electron laser", *Rev. Sci. Instrum.* **83**, 085117 (2012).

P. Zhang, *Beam Diagnostics in Superconducting Accelerating Cavities*,
Springer Theses, DOI: 10.1007/978-3-319-00759-5,

- P. Zhang, N. Baboi, R. M. Jones and N. Eddy, "Resolution study of higher-order-mode-based beam position diagnostics using custom-built electronics in strongly coupled 3.9 GHz multi-cavity accelerating module", 2012 *Journal of Instrumentation* **7** P11016.
- P. Zhang, N. Baboi and R.M. Jones,"Statistical methods for transverse beam position diagnostics with higher order modes in third harmonic 3.9 GHz superconducting accelerating cavities at FLASH", *Nuclear Instruments & Methods In Physics Research Section A: Accelerators, Spectrometers, Detectors and Associated Equipment* 2012, DOI: 10.1016/j.nima.2012.11.057.

Conference Talks

- `March 2013` Invited talk on *RFTech 4th Workshop* (Annecy, France), "Beam-excited higher order modes as a beam diagnostics tool in 1.3 GHz and 3.9 GHz superconducting accelerating cavities at FLASH".
- `June 2012` Invited plenary talk on *International Workshop on Higher-Order-Mode Diagnostics and Suppression in Superconducting Cavities* (Daresbury, UK), "A study of beam position diagnostics with HOM in the 3.9 GHz cavities at FLASH".
- `September 2011` Plenary talk on *2nd International Particle Accelerator Conference* (San Sebastian, Spain), "Study of Beam Diagnostics with Trapped Modes in 3rd Harmonic SC Cavities at FLASH".